Etheric Energy Exchange

Patti McTier

MILTON & HUGO L.L.C.
4407 Park Ave., Suite 5
Union City, NJ 07087, USA

Website: *www.miltonandhugo.com*
Hotline: 1-888-778-0033
Email: *info@miltonandhugo.com*

Ordering Information:
Quantity sales. Special discounts are granted to corporations, associations, and other organizations. For more information on these discounts, please reach out to the publisher using the contact information provided above.

Library of Congress Control Number: 2024910352
ISBN-13: 979-8-89285-095-7   [Paperback Edition]
979-8-89285-096-4   [Digital Edition]

Rev. date: 06/12/2024

Table of Contents

FOREWORD

The Law of Attraction is a force to be reckoned with, and for those who prefer scriptures, "As a man thinketh, so is he," "ask and it shall be given," "be transformed by the renewing of your mind," and, "faith is the substance of things hoped for, the evidence of things not seen," makes for a powerful combination.

Possession of what you seek in life requires more than simply running on the power of positive thought, though the Art of Allowing is certainly helpful.

The Law, at some point, requires action. We can't just bury ourselves under blankets, daydreaming for the best, and then magically our desires are made manifest. To quote the late singer, Rich Mullins, "Faith without works is like a song you can't sing, it's about as useless as a screen door on a submarine."

The object of this book is to help you understand where the two worlds meet.

My hope for you, the reader, is that this book challenges you to look at your life, as it has challenged me. Why do you believe the things that you believe? Are you hanging onto a belief out of fear? What if you could have a connection with your Creator, with whom alignment is a must, that's based on more than somebody's rules?

I hope you begin to understand the power of your thoughts and the things that you feed off of. Thoughts become words, words become actions. You must take responsibility for your mind, and thus, take responsibility for your life.

Ashley L. Franklin

This book contains a series of concepts to think about. Each chapter breaks down the concept in a way you can process and use the tips available to you to get the outcome you want or to better help you understand the mechanics between the physical world and Etheric realm. You will learn how energy flows through you, interacts with you, and how you can use this energy to draw in what you want. It is not all airy fairy. I'll explain how to use vibration and resonance and how to be observant of what the Universe is bringing to you. Each concept is a lesson in how to understand our reality. How can we pull from and utilize the limitless energy of the Ether and to get what we want out of life while not placing an imbalance of karma or taking from others? This is the science behind the Law of Attraction.

Definitions

Chi—is the life energy that exists in the human body and interacts with the Ether through the creation energy of Kundalini and the Chakra System.

Decibel—the pressure of sound; 85 Db is loud traffic.

Hertz (HZ)—the number of cycles of the frequency per second. One hertz is one wave passing a fixed point in one second. One hertz is equal to sixty vibrations per minute. An elephant communicates between 1–20Hz. A human hears between 500 and 2,000Hz.

Vibration—being able to physically tell or emotionally tell with your body the frequency of your surroundings (earthquake) or emotions (of yourself or others). To cause something to shake, in the Etheric it is the energy oscillating

Frequency—vibration physically measurable in hertz. This is what we detect as sound, either physical vibration from earthquakes or sound vibration from noise/music. This vibration then moves Etheric (emotion) or physical matter by moving through the person, object, Earth, the Universe. The phrase God spoke the world into being means the frequency/vibration rearranged the smallest particles (photons) to form the largest planets. We can rearrange objects now on sound plates. Different frequencies create different patterns.

Moving heavy objects through vibration: https://youtu.be/-2YXX7WXNKY?si=Rb3XECVL7qipeFG1 Experiments with sound: https://youtu.be/rYrdiQckGhw?si=tqiOrMdgmitADJfj

Water and sound: https://youtu.be/uENITui5_jU?si=ysByGXQoYlGsECLx

Resonance—is living/being in vibrational harmony with a person, place, experience, where the two frequencies (yourself and the other person, place, thing) make you happy, euphoric, joyful, and typically raises your frequency. Resonance can create healing of your emotional or physical state. Magnetism is a state of resonance and is either in complete resonance (attracts) or not (repels). As far as I know, there is no neutral magnetism. A metal can either be magnetic or not. It either attracts or repels another magnetic object. Ideas can be magnetic or not. Do you like the thought of money? If it is repulsive to you, you won't attract it. Opposites attract. Funny how we say, "Like goes to like but opposites attract." What I'm saying is that resonance brings attraction either vibrationally or magnetically. Objects are held at a certain point in time space and space time even though it doesn't look like there is anything acting upon the object. Resonance is the push and pull of keeping an object in place or at a certain distance from another object. Think stars, planets, people you like or people you keep a healthy distance from.

Space-time—*where* you are in the Universe at a specific moment (usually the present, some will say all that exists is the present).

Time-space—*when* you are at a place in time. For example, you are at a restaurant at 5:42 p.m. EST. Your friend went to that exact same restaurant and sat in the exact same seat at 7:05 p.m. EST. You were in the same space, just not the same time.

Chakras—energy vibrations moving through your physical and Etherical body. This starts from the ground up: root (red), sacral (orange), solar plexus (yellow), heart (green), throat (blue), mind (violet), Ethereal (violet/white). The energy also comes from the top down. This is bringing the Ethereal (ideas, dreams) into the physical. If you are reading this and feel repulsed by the thought of chakras and how energy flows through your body, who taught you to feel this way? We will explore these triggering emotions from words throughout the concepts.

Realms—different place, same time; same place, different time. One of the two must be the same. Once you know the exact similarity, a portal can be made of that frequency. Even if a portal is made, your body/spirit cannot pass through it until your frequency matches at least that resonance or within one, dimensional vibrations. If you are too low, you cannot pass through. If you are too high, you can pass through but will not be able to interact with that world, only witness it.

Dimensions (3D world, 4D ascension, 5D energy body)—as your physical body and Etherical spirit progresses, your vibration/frequency changes to match the realm you are in. Eventually, you get to a point where a body can no longer contain your energy and is no longer needed.

Density—vibration cannot flow through a dense object, thus the vibration is low. The vibration of sound travels well through water and air but not as well or at all through rock or the earth. Hence, the saying, "Are you dense?" Meaning, are you not understanding the information (vibration) given to you?

Plane of Existence—same place, same time, same vibration

Mantra—in Hindu, it means to physically manifest using sound. This is why you have positive mantras stuck on your mirror in the bathroom, vehicle, at work, or wherever you are wanting to remind yourself to think positive.

Money—energy in physical form that allows for a physically measurable/quantifiable amount of energy exchange; getting paid for your work

Ethereal field (aura)—energy that emanates from the body and moves out throughout the physical and Etherical space, creating a field in which your mind gathers information and your energy interacts with others in the physical and Etheric space.

Law of Attraction—you attract the same type of energy you are emitting.

Intuition—your mind processing data gathered from the ether.

Energy—the method by which each atom, photon, and/or particle interacts with one another throughout the multiverse, dimensions, and time.

Quantum biology—the field of study that investigates processes in living organisms that cannot be accurately described by the classical laws of physics.

Quantum entanglement—the phenomenon that explains how two particles, no matter how small, are linked, no matter the distance or time separation. As long as the two particles are linked via DNA or other physical matter or emotionally linked via Etheric Energy Exchange, the link cannot be severed.

Ether—the layer of energy in and around our bodies that extends through, around, and out to infinity. Ether is the energetic space that flows through all the Universe/multiverse. It is the web/grid/field that energy (some may call it plasma), information, ideas, and love flows through. It is the space in which Source/Creator/God exists. It is not a physical, tangible place in terms of a 3D world. It is vibrations that exist through all dimensions and planes of existence. It is the glue that holds all of creation together. It is the field. It is the place of spirit, the space of intuition and guidance, the space of ideas and memories. At some points in the book I use the words Ether, Universe, Creator and Source interchangeably. I realize the Universe is one verse out of all creation and the Etheric Field is all encompassing. I also understand that Creator, Source and God can mean different things to different people; however, the concept itself is very similar if not the same.

Field—a plane of existence that holds all of the energies and frequencies needed for life/existence in all forms.

Akashic Records- information contained in the Etheric Field. These records contain information on creation, our past lives, and the lives of all of the collective existing on Earth and in other realms and dimensions. Think of it as information stored in "the cloud." The Ether Net, connects the information stored in the cloud to all of the individuals in the collective through their spirit. Once your vibration is in resonance with that of the Etheric Field, (like tuning in a radio), you will be able to access the Akashic Records. In Sanskrit Akashic means "Ethers or that which holds all." In Buddhism Akasha is the fifth element, the source of reality.

Scalar Field—is the resonant field of the Ether. Vibrations effect the cells of living organisms, molecules and particles, to help create the blueprint of life and ensure that each program is running optimally and as designed.

Preface
Lost Books, Lost History

Something we all have to unlearn, research, and discover for ourselves is the history of humanity, the history of who we are as an individual and as a collective. If you ask the question to yourself, "who gets to write history?" it's the side who won the war, the side that has more power or influence over the other. Does the side in power always tell the truth? We know that answer is no, yet what we're taught in school is based on the side that had more power and influence.

The information we are taught is decided by a group that sees humanity as expendable, common, and only worth what we can produce for our short lifetime. For example, speculation and research has it that the Christian Bible originally had 777 books. What exactly is the Bible? It was a compilation put together by Pope Constantine in the late 390s. Monks were told what to keep and what to discard based on the Council of Nicaea. Who were the council members comprised of? What type of men? What was their agenda?

We can't be sure of the exact number based on the fact the Bible itself is a collection of works comprised of stories people wrote from secondhand, or even farther passed down knowledge of events.

The Dead Sea Scrolls, the Copper Scrolls, the Emerald Tablets, the Pyrgi Tablets are all examples of works that have yet to be fully decoded and understood. My point in this is, there are mentions of scriptures or writing referenced within this book that you may not have heard of. At that point, please do your own research. This book is about self-discovery and exploring the world and Universe around you. You can only do this research yourself. Quit being a human that is told how the world is, how to feel, what to think, what to do. Be your own sovereign being. Think for yourself and be in control of your own emotions.

Suggested Writings and Areas of Study:

The Magdalene Manuscript

The Gospel of the Holy Twelve

The Gospel of Q

The Apocryphon of John

The Gospel of Timothy

The Alchemy of Horas

The Emerald Tablets

Kundalini Rising

The Ethiopian Bible

Sumerian Texts

Astrology

Numerology

Tarot

The Silva Mind Control Method by Jose Silva

Know Your Own Mind and How to Foresee and Control Your Future, by Harold Sherman

Royal Raymond Rife and Rife Technology

Stalking the Wild Pendulum, by Itzhak Bentov

The Body Electric, by Robert O Becker MD and Gary Selden

The Science of Getting Rich, by Wallace D Wattles

A Guide To Leylines Earth Energies Nodes & Large Vortexes, by Rory Duff

The Quickening, by Stuart Wilde

The Farmer's Almanac

Return of The Divine Sophia, by Tricia McCannon

Concept 1

What Is the Ether?

What is the Ether? How do we use the energy around us to live an abundant life? The Ether is the layer of energy in and around our bodies. Most people think of the Etheric field as the place where angels or spiritual energy resides. After all, we are spiritual/energy beings on a human journey. The Etheric Energy Field contains universal knowledge, ideas, and thoughts of the past, present, and future. Time does not exist in the Ether. We utilize our connection to the Ether to pull our thoughts, dreams, and goals into existence. The Ether is where our mind exists. We use our brain to generate the electrical activity that becomes the conduit for the electrical and magnetic energy that resonates with the Ether.

The Ether is all around us. We pull the Etheric energy into ourselves, into our physical body, and create an action. If we do not act on our thoughts or ideas, nothing will ever happen. We can pray or meditate to boost our Etheric energy, but if we never act on our thoughts, or even speak our words to share our thoughts with others, nothing will ever progress. Your idea, your dream, will never come into being if you never speak it out loud. You may want to start a business or write a book. You may want to travel or build a house. If you don't start it in the physical world, you will keep dreaming about it, and that's all it will ever be—a dream.

People like to call this concept of pulling from the Etheric to the physical *manifesting* by using the Universal Law of Attraction. Many people are confused as to what manifesting is. Manifesting is placing your intent out of your mind and into the physical world. The Ether/Universe/Creator can then work to rearrange energy for opportunities to happen through the vibration and resonance between you and other objects or emotions you want from the physical realm. For example, you want to find love. You put this intention out into the world by speaking about it. Then one day you randomly meet someone at a store you've never been to before, and you end up going on a date, and that interaction turns into dating and then love. Maybe you're at a gas station and you bump into the owner of a company where you're trying to get hired, or you stumble across the piece of information you need to get your health back while scrolling through social media. *But*—and it's a big but—if you never have any action behind all that energy the Universe has rearranged for you, nothing will happen, nothing will manifest. If you want a mate but continue to sit at home watching TV, your serendipitous moment will not happen. If you never step up and say "hi" to the person next to you at a gas station or in the coffee line, your opportunity will pass you by. The Universe did its part...did you? This is the Universal Law of Action. You can *attract*, but if you if you don't *act* and continue to act in a manner congruent with what you want to attract, the Ether will stop trying to manifest opportunities for you.

8

If you act, the Ether must respond. How do we know? If you want to lose weight, you must put your thoughts toward that action. If you complete those actions, eating right and exercising, you will manifest what you want. The Universe will do its part to bring you a workout partner, a trainer, or maybe a discount special at your local gym. The same goes for business. If you continue making sales calls, getting out and meeting people, eventually you will start creating a sales pipeline, converting sales, and money exchange (an exchange of energy) will occur. But if you never workout or never make a call, nothing will happen on its own. Our action causes a reaction in the Ether. Not an equal and opposite action, but a pulling attraction that will match your energetic vibration.

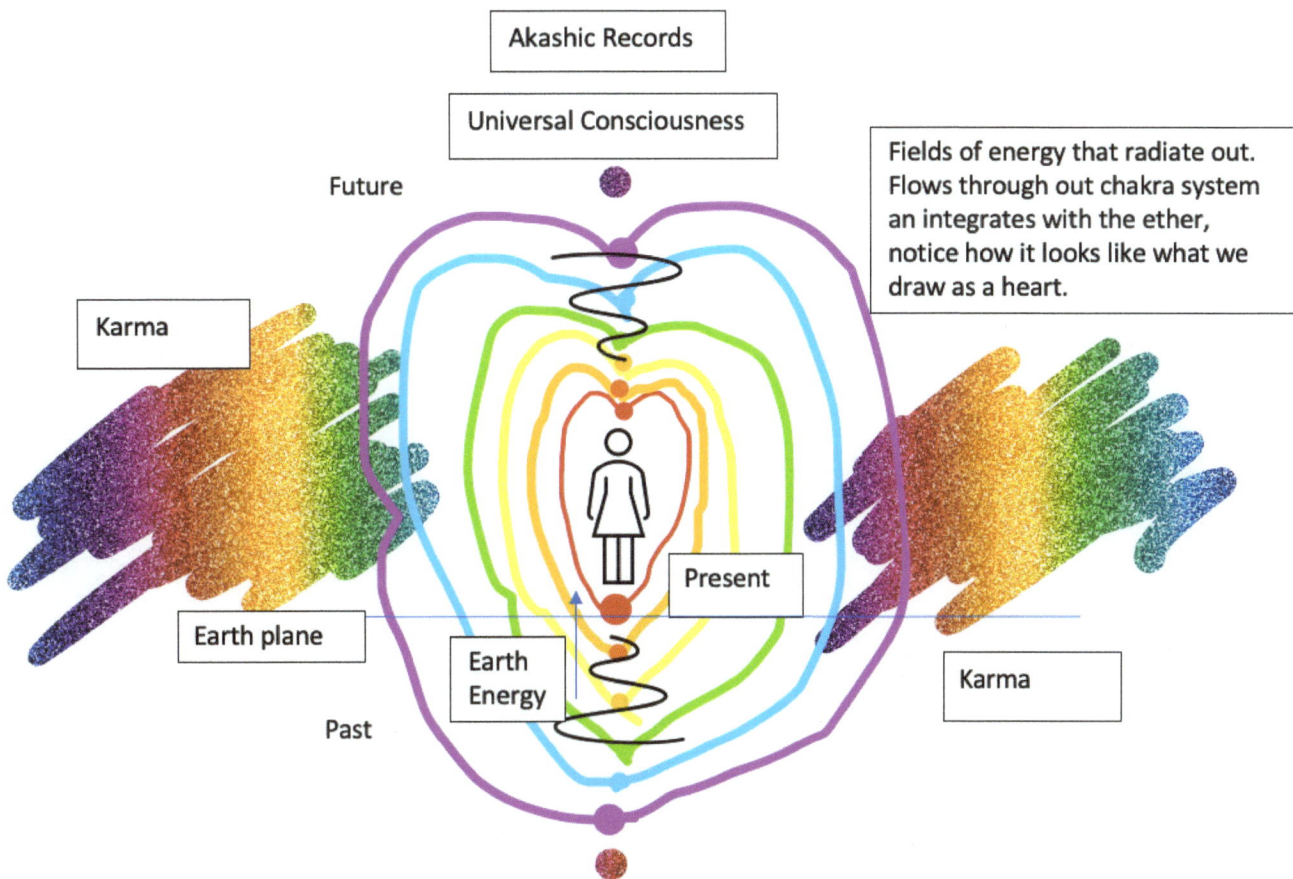

Where we stand on the earth plane is obviously the core of our physical existence. We stand on the earth, gain energy from the earth, and bring this energy up through our bodies. This is called *grounding*. The magnetic energy from our own heart is in perfect resonance with the frequency of the earth. This is why being out in nature is so calming. This is why being next to a waterfall or taking a shower at the end of the day and having all those negative ions wash over our energy bodies washes off all the emotions from the stress of our day. Our chakra system runs through our bodies connecting the physical earth with the Etheric field.

The Akashic records, the information of creation, surrounds us. What some call Universal Consciousness is the interaction of the energy fields of all humans and all beings in existence interacting with one another.

Universal Consciousness exists in the Ether. It is how we are a collective unit but separated by our own auras as physical bodies. This is the oneness of the Creator; and how, through quantum entanglement, we can know when a loved one needs us or how we can feel an impending event for the greater humanity, such as a war, anticipation over an election, or a weather event. It has been quite a while since there has been a global event of great joy, but it does happen. Through the Akashic Field, this is how we find our soul family. Have you ever met someone for the first time, but felt like you've known them forever? This feeling is your memory resonating with the other person's aura or energy field. Your energies know each other, even though your mind doesn't remember.

Concept 2

Occupying the Ether

Energy Exchange is measurable, quantifiable, and is felt physically and emotionally. Sometimes you can see it, but do you see the energy before you hit the light switch and the bulb lights up? Do you see the energy coursing through the wires? Do you see lightning before it strikes? No. The energy is there. You can feel static in the air, a "vibe" in a room, or you can pick up on a person's emotions. We also get feelings from ideas. What resonates with you?

In the Ether, there is space-time. You are occupying a space at a particular time where you are currently existing. In our world, we have places we want to be, levels we want to progress to, and become the person we hope to be. When we get stuck/stagnant in one space for too long, we could energetically be occupying someone else's space.

For example, as a teenager, you get a job a local restaurant. You never acquire any more skills. You drop out of high school and remain working at a restaurant. Your potential was much greater than what you fulfilled. You are now occupying someone else's space. When human beings do not pursue the greatest version of themselves and they linger in a space they shouldn't have, they then begin to occupy the space that should have gone to someone else. The Universe will reroute and take others around your decisions eventually. This web of decisions where people are interacting with others, whether they mean to or not, creates the collective consciousness. This is sometimes called the butterfly effect, where the smallest effect by one can create a dramatic difference on another. This happens because of individual decisions within the Collective Consciousness. This is not a hive mentality, like artificial intelligence or bees, but a collective where we are all connected by thoughts, feelings, and magnetic fields that surround each of us, this planet and the multiverse, through quantum entanglement. We will get more into quantum entanglement later.

In another example, say you volunteer for a local charity. You learn the dynamics of volunteering. You feel the emotion of helping others. You work together with like-minded people to make a difference in your community. You continue volunteering your time and skills in the same role and you never take on any additional responsibility. You never go out into the world and teach others about what you have learned or share your experience with others, or taught others how to get involved. It is great that you feel fulfilled and you are doing a service to your community, but you should transition to holding space for someone else to walk into, to make others feel welcome to join you in your efforts.

To reinvent yourself is to grow your energy self/your spirit within the Ether, to allow more Etheric energy to flow through you than your body/container/vessel is capable of holding on its own, to become a conduit for change and exchange the energy of feelings and thoughts with others.

There are times when we are waiting for other people to catch up to us or people are waiting for us to provide a safe place for them to have an experience. This is called holding space for someone else. The most relatable way is holding a safe space for children to grow up by having a home that is warm, nurturing, loving, and forgiving, as well as having safe boundaries. Another way is when you are working with other people in a guiding or mentoring capacity. You create a safe space for learning to take place, for people to ask questions and make choices without judgment. I'm not saying that these choices are free of consequences, but you are there to prevent your mentee from doing any major damage to themselves physically, mentally, or spiritually. Mentally and spiritually, you are holding a space and helping the person ask questions of themselves. Perhaps you have insight as to how others perceive them, making them self-aware. Are they too closed off? Maybe they're too open and not setting good boundaries? They could be unable to communicate their thoughts and emotions in a productive way. Do they have the information they need to make choices in their highest and best good? You guide the person by offering advice, but that is all you can do. You hold space for them long enough for them to come to a conclusion on their own. As a guide, you must brace yourself because your students may not choose as you would, but that's all you can do—create that space.

Beware of being drained as a light worker. You cannot hold space for someone if they are syphoning off your energy while attempting to build their own. This can happen in close relationships where the light worker is trying to help but their energy is being drained while trying to help the other person. For example, waiting for someone to decide if they want to be in a relationship with you or waiting for your partner to gain enough skills to progress into a career they want, yet no progress or transformation is happening. They begin to live off your energy, take advantage of you, or do not appreciate what you do for them. If this happens, you need to recognize it and distance yourself from people who are energy thieves. Again, it may not be the outcome you hoped for, but you held space as long as you could before you had to reset your own boundary.

This is the beginnings of setting up a barrier, your "bubble". Like goes to like. Based on your vibration.

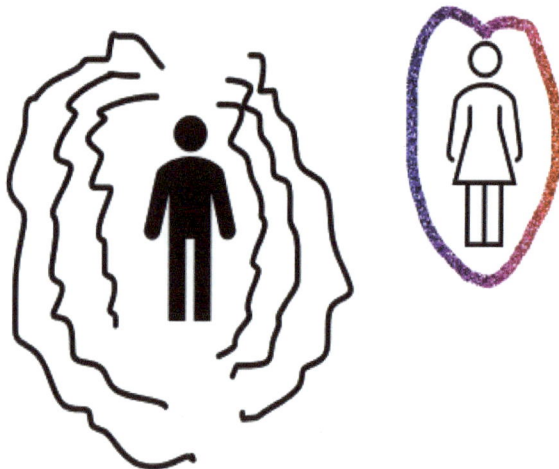

Aggravating energy attracts other aggravating people. Insulate yourself from negative vibrations by increasing the positivity of your aura. Think happy thoughts, be around positive people and be aware of energy theives.

What Resonates with You

Resonance is a vibration, a frequency, and when we resonate with an idea or a person, it means their vibration matches our own. This feeling is what gives us the confirmation to move forward and do something: start a relationship, to pursue the idea, or take the job. Resonance is how we choose an area of study; however, some people did not choose their education. For example all children in the US are sent to school for the same curriculum. While a base study of all subjects provides us all with a sampling of the world, why do we gravitate toward one area of study? Also, school only teaches five core subjects. How do we know what else is in the world if we are limited to five narrowly taught subjects? If we gravitate toward music or science, why? Why is it fun to read books? It is because the author is resonating with us? Or perhaps you detest reading and would rather watch a video or make something with your hands. What we learn and how we learn is important to the growth of the human soul. In case you thought you were growing your brain, not so. Our bodies grow old and die, along with all our organs including our brains. Since energy cannot be created or destroyed, it is our mind—our thoughts and memories, our soul, that we take with us.

How do we know a liar? Many people get in bad relationships because they cannot feel the signs of a narcissist. They ignore their feelings. We are taught from a young age to ignore our intuition, our gut feelings.

While learning, we are interacting with the Ether and Universal Law. We are basing our learning subconsciously with our engrained biological hardware of our DNA. What resonates certainly is not a common core curriculum. You can take what is taught or you can take what resonates. Also, our resonance changes the more we learn. We may have a certain way of thinking; but as we learn, by being exposed to all different perceptions and subjects, our frequency, and therefore what resonates, will change.

When we talk about resonating, we are talking about how we are all part of one collective consciousness—the Ether.

During this time of the woke versus the awakened, I feel the need to talk about the 100th Monkey Effect. *Wikipedia* describes it as "a hypothetical phenomenon in which a new behavior or idea is spread rapidly by *unexplained means* from one group to all related groups, once a critical number of members of one group exhibit the new behavior or acknowledge the new idea." The *unexplained means* is the passing of the idea through the Etheric Energy Field surrounding the group i.e. the Collective Consciousness. This is the energy of the Ether and is measured by the Schumann Resonance. Many mainstream "scientists" have tried to discount the idea of a Collective Consciousness, yet many other studies have shown that if 10 percent of a population can be persuaded to believe a certain narrative, the opposing narrative begins to lose the resonance of the whole. Now think, if 50 percent, 80 percent, or more believe in a certain idea, does this make it true?

The point of understanding resonance is paying attention to our feelings. Acknowledge those feelings and then act accordingly to move us in the direction we want to go. Pay attention to the pangs in your gut telling you something is wrong. Don't stay with someone out of niceties or obligations if the person doesn't resonate. Don't stay at a job that quite literally drains the life out of you.

Your frequency goes out into the ether and begins to pull toward you those that resonate with you, that have similar ideas, business partners, love partners, your "tribe." It can also help manifest where you want to live, how you want to design your life.

Concept 3

What Does the Interaction with the Ether Look Like?

You interact with the Ether when you sleep, during the moments of waking up, or when you're daydreaming. This interaction happens during what is called the Theta state of brainwaves. People try and put themselves into a state of meditation or relaxation to see what ideas come to them. By developing your intuition, your discernment, and having clear wants and desires, the Ether can begin to bring you energy that resonates with what you are trying to create, like a magnet.

If you think, *I want to attract money.* Great, but that's ambiguous. That's like going into a grocery store and saying I want food. There are all kinds of food stuffs in that store; fruits, carbs, proteins, junk food, maybe even something you're deathly allergic to. That is how people get frustrated. "God, I wanted a good mate. Why did you bring me a lazy, energy-draining, unattractive person?!" Well, when that person was presented, did you have your boundaries set? Did you use your discernment? Did you ask the right questions? "Is this the person I want?" Then tell the Universe, "No, I want a kind person with a good work ethic." So you said no to person number one. When you acted and said no, the Universe refined its search for you and presented you with another option (choice). "Okay, here is a person that is kind and hardworking, but they don't want kids and I do, or maybe they do not have the same goals as me. No." Okay, Universe goes out and searches again. Throughout this process, you are putting out good vibes, your authentic vibes, and you're using your discernment. You're being clear with what you want and being honest if that person resonates with you or not. This is why it is so important to be your authentic self at all times. When we want one thing, but portray another, the Universe gets confused about what to bring you, so it either brings you nothing because nothing is attracted or brings you the "wrong things" because that is the vibe you're putting out.

If you are wanting to build a business, what actions can you take right now to start on that path? Be congruent. Make choices that will continue to put you on the path to your goals and passions. If you are presented with a choice, make the one that puts you on the path you want to be on. As long as you are clear and act, the Universe has to respond in kind to your thoughts and actions by bringing you opportunities (aka choices). If you want a relationship, or a vacation, a house, anything, you must maintain the vibration that will bring it to you.

What type of person owns a home? The answer is a person who can pay for a home. A person who does activities consistently that will create the environment for a home to happen.

You want a vacation. Do you think about where you want to go? Do you save your money to budget in order to get there? Do you *tell* people about where you want to go? Remember writing and speaking into the physical realm helps to bring your ideas and dreams into reality.

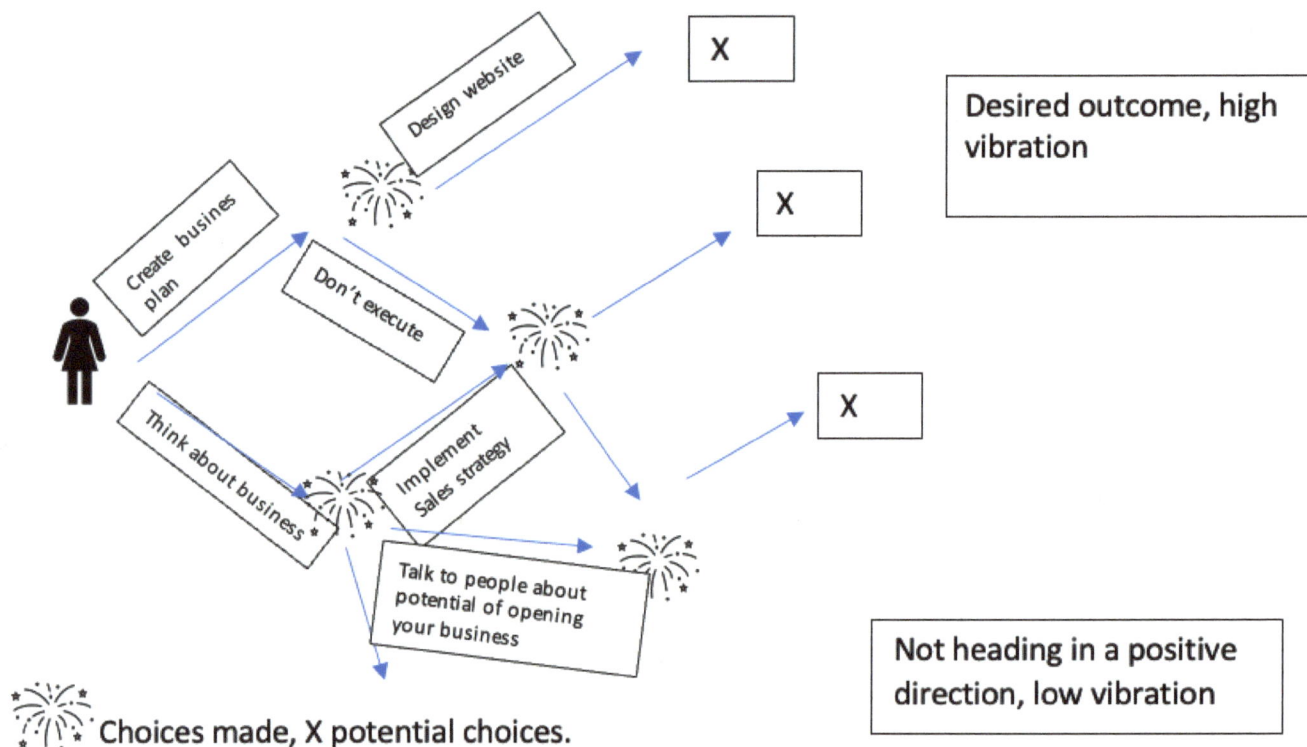

Design website

Create business plan

Don't execute

Think about business

Implement Sales strategy

Talk to people about potential of opening your business

X

X

X

Desired outcome, high vibration

Not heading in a positive direction, low vibration

Choices made, X potential choices.

How Do We Interact with the Ether?

This is basically what Unified Field Theory and Quantum Physics has tried to explain, and humanity is continuing to discover the "new" (but really ancient and esoteric knowledge) information about the Universe.

Position your body as the center of your world, the center of your existence. You are physically standing in the present. To your left is your future; to the right is your past. Time passes through you as a function of gravity pressing from the top down. Surrounding you is the Ether. The Ether contains energy, Universal Consciousness, knowledge from past lives, ideas of others, your ideas, as well as emotional energy. This energy of memories, thoughts, and ideas is known as the *Akashic records, Universal Knowing, Universal Consciousness or Spirit*. The reason why Akashic records applies here is that we actually live in a multiverse, not just a "uni" verse. More like verses of a song or a frequency, multiple resonate tones play at once to create a tapestry of physical manifestation through tones, frequency, and resonance.

There are three fields needed to construct our physical reality: (1) Kundalini frequency of creation, (2) the Magnetic Field, and (3) our energy body/Chakra system (power system channeling energy) including the twelve meridians (the circuit breakers and wiring of your body, literally wired into your nervous system).

In the physical realm, we also have time. Time energy spirals through us. It looks like a spring. The energy wraps around our body, encompasses us, and anchors our physical body to a specific space-time. We are able to manifest our thoughts into the physical by drawing in creation energy also known as *Kundalini* energy from the Ether. So time looks like a spring; *Kundalini looks like the double helix* of DNA; and the *Magnetic field* is a toroidal field and looks like concentric strings coming from the top of your head, out in a heart type or circular fashion, down through the Earth, and up again to your feet.

The energy of all three fields moves both ways, from the top of your head, the Crown Chakra, to the bottom of your feet and into the earth, way below the root Chakra. This is basically the trinity of energy. You need all three to hold your existence in the physical (earth), mental (Ether), and spiritual (multiverse). All three fields interact with your body and your thoughts. I have to note, when I see the energetic toroidal field drawn around a human body, it stops at the root Chakra. If you are living on earth (which I think we all are), the bottom of the field is what connects you to the earth; and actually when you expand your aura, your energy field, it can drive down deep into the earth and expands above you and out around you. You always want to expand your sphere of influence. This is literally/physically how it's done.

Mind, body, and spirit are not separate while being an active embodied member of the third dimension (human). The mind is connected to the body through the brain, much like a software is to hardware. It can be disrupted if the brain (hardware) is disrupted by poisons, toxins, viruses, frequency interference, and programming.

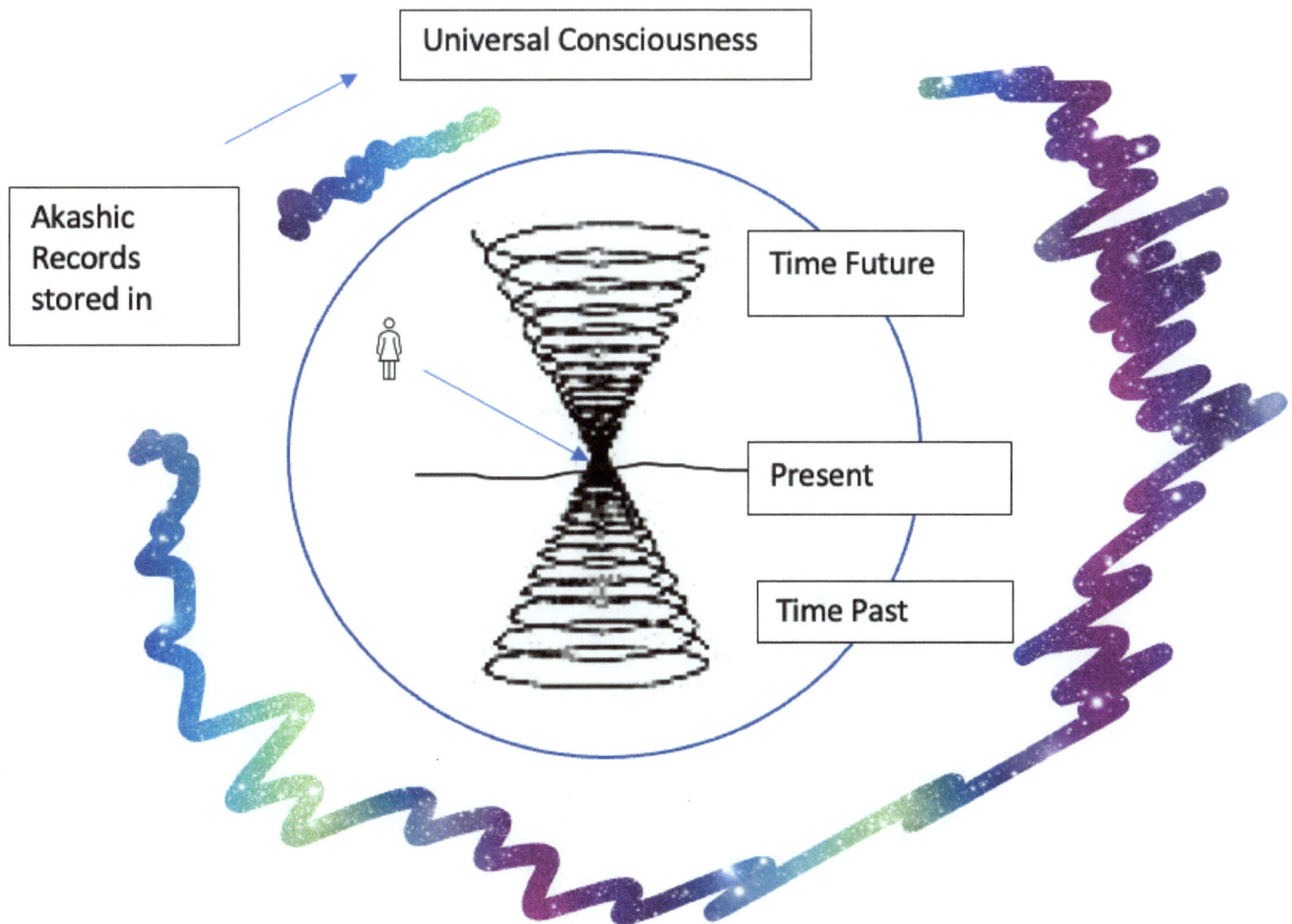

Kundalini etheric energy body

Magnetic field physical body interaction the interaction between the physical and etheric

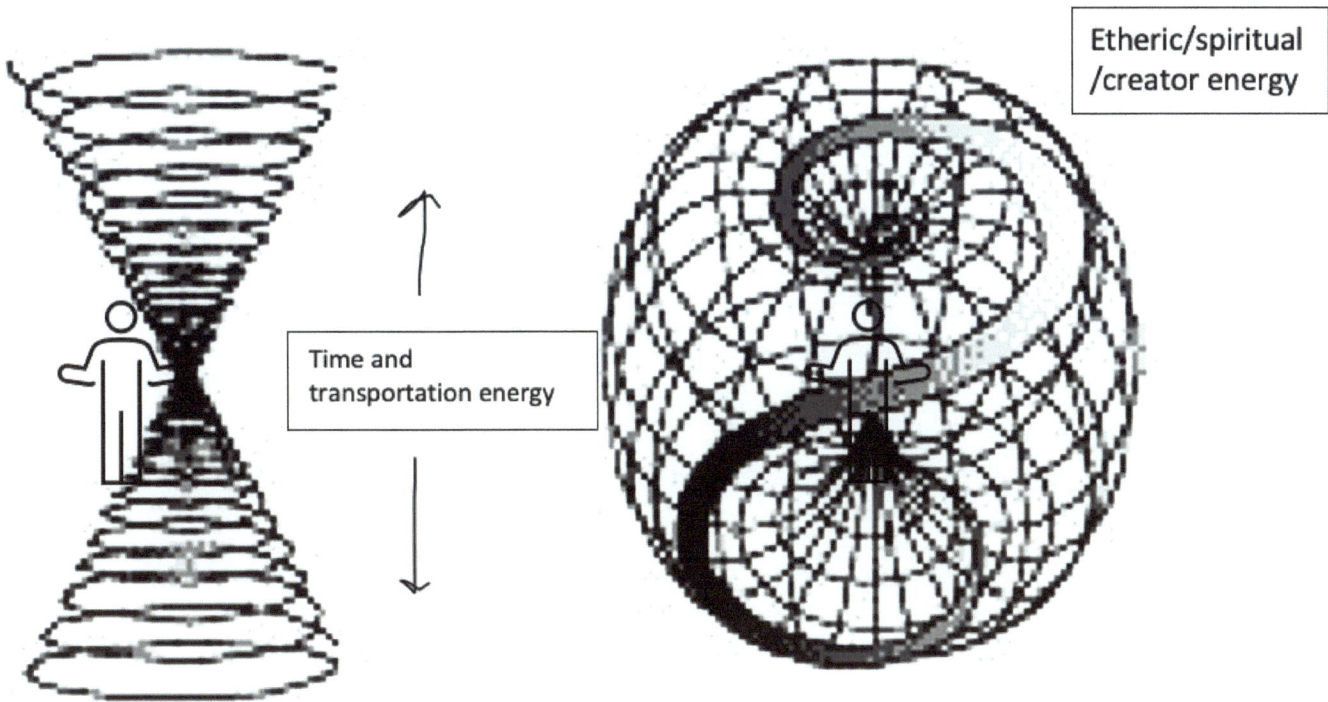

Time and transportation energy

Etheric/spiritual /creator energy

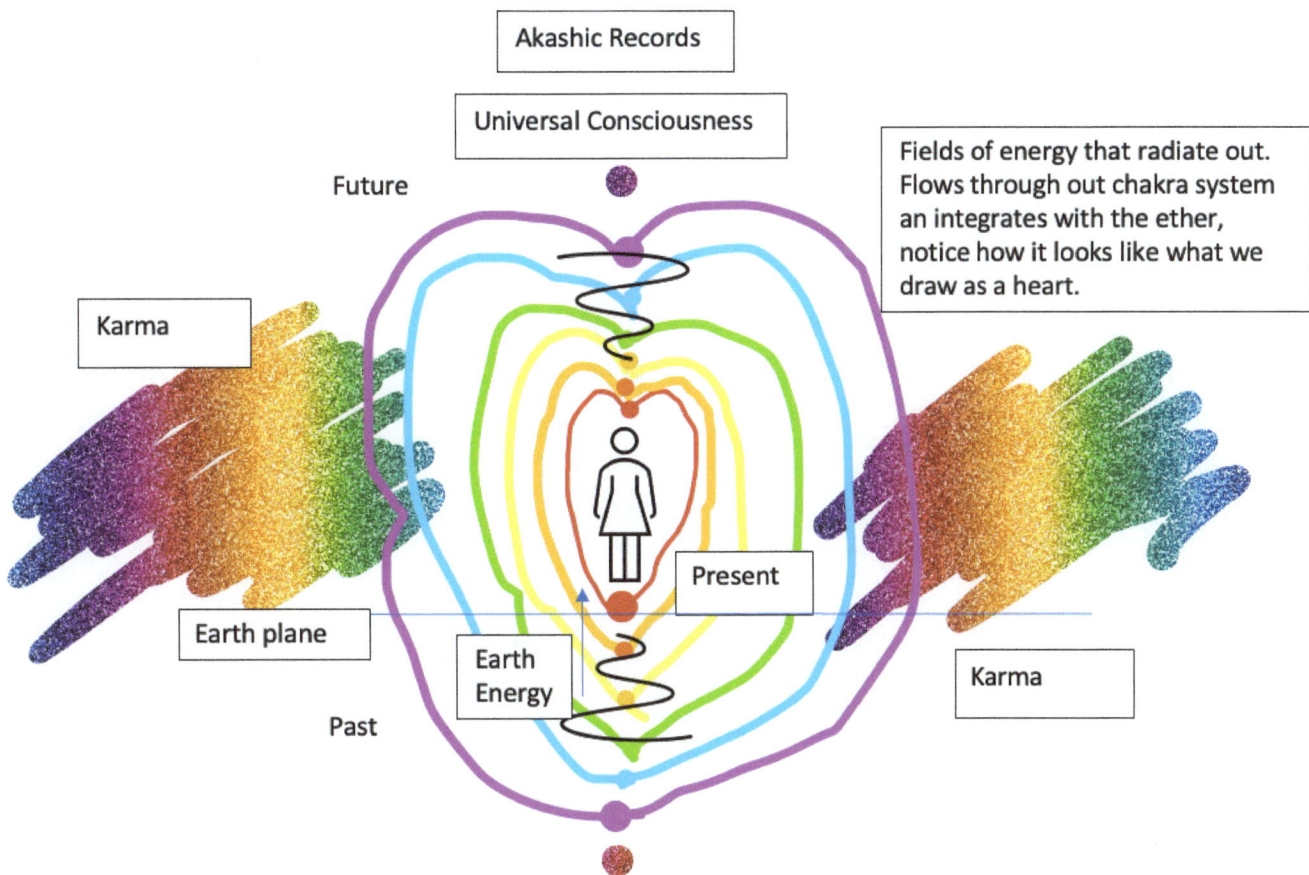

Akashic Records

Universal Consciousness

Future

Karma

Fields of energy that radiate out. Flows through out chakra system an integrates with the ether, notice how it looks like what we draw as a heart.

Present

Earth plane

Earth Energy

Karma

Past

Concept 4
Energy Exchange

Energy exchange is measurable, quantifiable, and is felt physically and emotionally. Remember when we discussed electricity and the "vibes" earlier? Sometimes you can see energy, but most of the time you can't. Do you see the electricity pulsing through the wires in the wall before you hit the light switch, before the bulb lit up? Do you see lightning before it strikes? No. The energy is there. You can feel static in the air, a "vibe" in the room, a feeling from a person. We also get feelings from ideas. What resonates with you? How does the idea make you feel?

Call the energy exchange what you want, love, gifts, work or passion. The energy within the exchange must balance. You can't have an unbalanced equation/interaction. The Universe does not work that way. The energy comes to, through, and from your body. Whether you believe in this unseen energy doesn't matter. Illustrated below, we all have energy in the form of love, gifts, work, and passion. If you are not creating from your passion, you cannot tap into creator energy and manifest more than what your body/vessel can produce. Growth occurs from learning new ideas or skills, thus growing your container. This will be explained more fully in concept 7.

Think of yourself as a battery. Your vessel can hold a certain amount of energy. You receive energy back from the harvest of your work, ideas, or your gifts; however, if you want to generate a higher output, there are several ways:

1. Work more, which will only work for a while: law of diminishing returns. You will have a higher output, but it will put a toll on the battery, your body. Your body's capacity will begin to hold less and less energy if you tax it too much.

2. A gift to boost your output. This can be a monetary gift from someone; investing in your business; or a gift of someone else's energy, working with you to help complete a task. This can also be in the form of a partnership or working with a group of people to accomplish a task.

3. You can gain a skill or education from absorbing other's energy, attending a class, or by sitting quietly/ meditating to figure out how to solve your problem. By having a greater skill set and more depth of knowledge, your value will increase and the quality of your output will increase; and thus, your pay should increase. Skills and knowledge increase the capacity of your vessel.

The phrase "work smarter not harder" seems to fit. To balance out the equation, "positive" or "negative" energy does not matter, meaning only units of energy matter in fulfilling the energy exchange. We humans

20

are the ones putting an emotion to that energy. The Ether manifests energy to fill the void in the gap between what we are paid (energy) and what our gifts/skills/thoughts are worth (also energy).

Here is another illustration of your vessel, your container, your body:

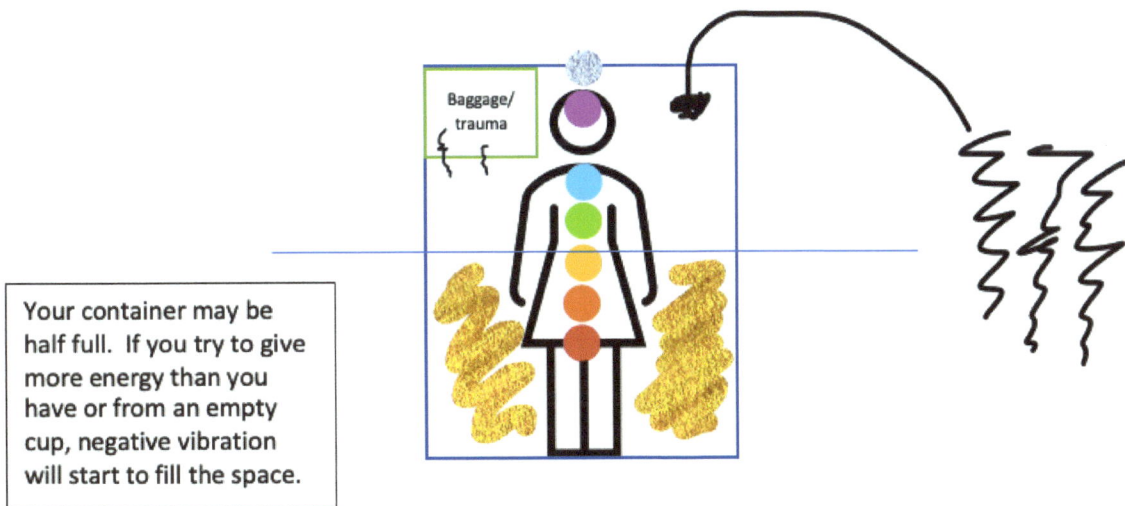

Baggage/trauma

Your container may be half full. If you try to give more energy than you have or from an empty cup, negative vibration will start to fill the space.

Your frequency goes out into the ether and begins to pull toward you those that resonate with you, that have similar ideas, business partners, love partners, your "tribe." It can also help manifest where you want to live, how you want to design your life.

The only way to increase our capacity as an energy being is to become a circuit rather than a battery. When we gain full expression of ourselves, living in our gifts and creating from our passion, our bodies become a circuit for energy to flow through. We no longer contain a finite amount of energy but are able to tap into the infinite energy that exists in the Ether.

How do we properly value our gifts/skills/thoughts? We establish what we need and what we want, not the monetary amount but the earthly needs and wants as well as emotional and spiritual.

Needs

Home: Home is not just "shelter" but where you want to live. If you've always been called to live near water, do it. If the mountains or a certain city has always appealed to you, then go. If you are not in your proper place, Source/God/Ether cannot give you what you want. You must be in the right place and around the right people, because you are pulling the energy from the Earth. You need this energy to replenish yourself. This is like a prayer spot or vortex where energy can flow into your body. Home also needs to be where you can accomplish your passion. For example you want to be surfing instructor, you would need to live near a beach. If you wanted to be a full-time gardener, you wouldn't choose to live in a harsh climate.

For me, I started feeling the need to move in 2018. How did I know I was supposed to move? I felt like the city my family lived in no longer fit us. It was becoming too crowded. Despite the fact we had many young children in the neighborhood, my boys were never invited over to play. In fact, there were so many kids it was sensory overload in the noisy lunchroom and gym to the point my younger one had to wear noise cancelling earmuffs to prevent from being overstimulated. When I wanted to go to the store, I had to make sure I was leaving at a time when traffic would be minimal or the trip would take three times as long.

I began to feel a huge push when my work was leading me out of state, to a tiny little town in Tennessee. We came up over a weekend and looked at a handful of houses, of which maybe one of them remotely appealed to us. I began searching on my phone and happened to find one on the opposite side of town from where we were looking, and it was perfect. A house on five acres with a view of the mountains and the lake. Ironically, the view is quite like a picture I would draw often as a child: Mountains with a lake at the bottom, grasses growing tall, and a lot of wildlife. We made the choice to move. I did an immense amount of cleaning to prepare our current house. There is no kind of purge quite like that of putting your house on the market. Our house sold in two days with a solid offer. The point of the story is, when you answer a calling, the gift the Universe is trying to give you could be nothing what you would have imagined for yourself.

You never know how your gifts from Source/Creator/God/the Universe are going to manifest. It may be exactly what you want, just not in a place you thought. Or it may come from a different angle than you ever thought possible. Astrological charts, such as your birth chart, can be very helpful when it comes to moves in your life, which areas of study you should pursue, and the types of people who are most likely to resonate with you. Basically, how to be in the right place at the right time. We will talk about synchronicities and how to identify them later in the book.

Clothes: Wear what feels comfortable. Wear colors that make you feel happy and confident. When you put your clothes on, you should feel like you are preparing for your purpose/passion.

Food: Food is as important as choosing your work and sharing your gifts. Honestly, I am still learning about this, and the human diet is a billion-dollar industry built around this sole necessary habit. We need to feed our physical body *and* our energy body. You can find what type of energy body you are by going to www.myhumandesign.com. To determine your energy body type, you can also use the doshas. There are several sources online. Eat as close to the earth as you can with as little ingredients as you can. And eat as much from the light as possible (fruits and veggies).

Love: Love is necessary. It's a need and a want. Some of us have felt love our whole lives: gifts from our parents, our friends, children, or partners. For some, love has been elusive or even abusive, something that has never manifested in the physical or something that has manifested in a distorted way. It's an idea, a thought that is stuck in the Ether. What do we need to manifest love? First, you must manifest love for yourself. You must recognize that you are your own person, that you are worthy and valued. It starts by knowing that the Creator, the Universe would never create without love. Negativity/evil can never create. It only contorts and manipulates what has already been created. You must find a way to reconnect to Source/God/the Universe by connecting to yourself. You are the conduit. Your body, mind ,and soul is the radio that by turning the dial and scanning the frequencies can tune in to the message you need to hear. This can be done through meditation, yoga, prayer, or getting out in nature. Sit by flowing water. Flowing water has a way of washing away negative energy. Ironically, it's negative ions that have a positive effect on our emotions. So find a place to sit at the beach, a flowing stream, a waterfall, or where water can rush over you. That's why we wash away our bad day down the drain in the shower or soak in a tub to relax.

Spirituality: Love and spirituality go together. Spirituality is the understanding that we are one component of a larger more complex energetic system. I hate to break it to the atheists, but their nonbelief in a greater energy or no source simply proves their ignorance to the information provided to us through the quantum. Quantum biology and quantum science in general, is the proof of a Creator or Central Source. Having a belief, appreciating creation, and taking care of yourself are all love and spirituality.

Purpose/Passion: What would you do if money did not exist? What would you do every day if you had everything taken care of, food, shelter, love? Who would you want to help? Women especially have an issue with this as they continually want to give from an empty cup. If your cup/vessel/body is not filled, then the energy exchange becomes imbalanced. What happens is your gifts become a drain on the other person receiving the gift. For example, you gift to your children your care, your love; however, if your body is physically tired, you're grouchy, you haven't spoken to another adult besides your spouse in weeks. You then begin to have an empty cup. Your love becomes obligation, a necessity, instead of enjoying little faces. I mean, how could you right now with all the screaming, poop, feeding, and mess literally everywhere you look. Your cup has become empty. You must reach for the Ether/Spirit/Source. You need to find energy outside of yourself for replenishment. Be in nature. Take a bath or a shower for more than two minutes if possible (slightly joking but not really to the moms). I am not saying to look outside of yourself to someone else, but looking for an energy outlet that you can plug into to recharge yourself.

What about a job that continually drains you? You wake up dreading going in. No one talks to you, or you have to put up with the public. Or you work from home, and you feel isolated. If something is creating an imbalance, identify where it is coming from. Is it something you can change? Where can you draw additional energy from to replenish yourself if you cannot change your current situation?

Going back to what would we do if we did not need money. This is usually our passion. I heard a quote, I can't remember where, it said, "You don't earn money to do what you want, you earn money doing what you want." Having this idea as a starting point can help you start to implement steps to getting you where you want to be and who you want to become.

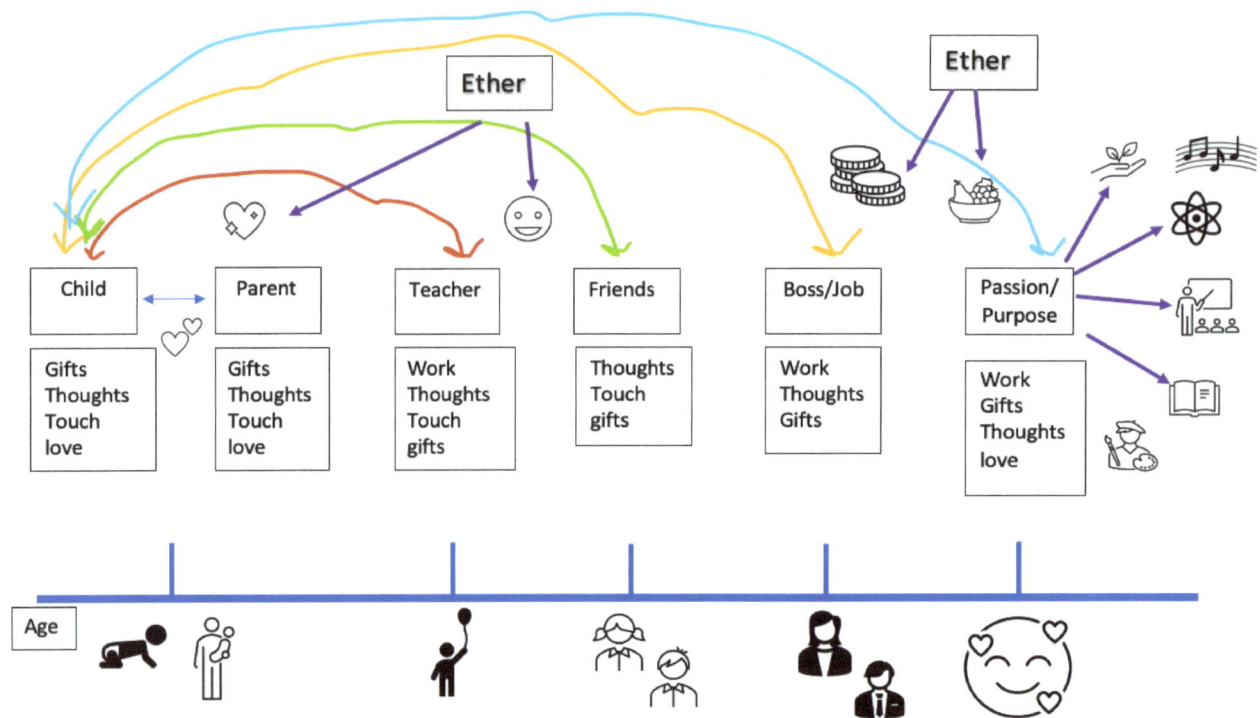

Wants:

What do we want to manifest? House(s), car(s), jewelry, a vacation, health, a social life, the ability to eat out whenever we feel like it and not having to worry about how much groceries cost. What are your nonnegotiables? This is not a dollar amount, but what you will and won't do for energy exchange. Syphoning of your energy should not be an exchange. Anything that makes you feel tired, angry, apathetic or lazy. Some people use the term *laziness*, but what if it's not laziness? What if you are genuinely tired of being where you are because it's a total energy suck?

Children being forced to sit still for six plus hours a day, not talking, not doing any kinetic learning; and if they throw a tantrum not wanting to go, it's their body's way of saying that energy is not in alignment with them. When we don't want to go to work because of how it makes us feel, it's not in alignment. We are choosing to live without joy and convincing ourselves it's okay, telling our children, it's okay. It is *not* okay.

None of our wants are bad. We have been programmed to think that money is bad, wealth is bad. Every time you think money is bad, replace the word "money" with "energy." Ultimately, money is a physical representation of an exchange of energy. If you don't have "wants" you don't have "goals," and you won't be able to focus and manifest on what you want or need.

Since we are all on this human journey, and none of this falls from the sky, what do our gifts/skills/thoughts allow us to work for and how do we best share our energy with the world? What do you feel called to do? Art, technology, helping animals, a better way of health care through light, vibration, frequencies, or developing a new way to travel through magnetic propulsion? What about growing food on a community scale without the need for toxic fertilizers? Your idea cannot be vague, such as own a successful business. Your idea in the Ether has to be able to connect to something, some object in the physical that has a resonance that can match yours. That's when the pull can start to happen!

Whatever you are drawn to do, figure out how can you take steps in that direction. The idea can start as a hobby, then grow from there. The point is to start something physical in the physical world. Move it from concept to creation, idea to implementation, etc.

You could need new skills. Often this is the case. You may work as an English teacher now, but you may want to become a nurse. What courses would you need to take, where can you take them, how much would it cost? Better yet, find someone who is a nurse and ask them about their experience. Is it really what you want to do? Is changing your profession actually what you want? Maybe you are drawn to helping others in a medical capacity; but you find that changing out bandages, surgeries, or needles is not something your stomach can handle. Maybe you choose dentistry, pharmacy, or naturopathy. You can always become an apprentice or shadow someone for a day. Take your friend to work day. LOL. I think exploring the idea without making monetary moves is a great first step in leading toward a decision. This is not just for young people. This is for anyone who wants to make a legitimate change in their life but cannot go full tilt into something new.

The value you put on your time is based on your skill, but also needs to be close to the value others have for your skill.

Don't sell out. Do not be dissuaded by even your closest loved ones. They are not given the same gift or calling as you. You must find your "tribe," people who have the same energy frequency as you. Have you ever felt drained after being around someone? Or maybe you find it difficult to have a conversation with someone (this can be family members). If the person does not resonate with you, you do not have to continue to engage with that person. Even if it's family, especially a parent or sibling, try and limit your exposure to the syphoning of your energy. It's okay to have boundaries to protect yourself.

Actions That Give You Energy versus Actions That Take Your Energy

Taking care of your child is sometimes draining (me laughing inside). But because of the love that you have for your child, you are able to continue to give, even if you feel drained. When you can take a break and relax, you draw from the Ether to bring your energy back, as in the water mentioned above.

It is the same for the work we choose. I have heard on many occasions that painters or coders will sit for hours on end, working their craft. They become so immersed in what they are doing, being in "the zone," that it is nothing for eight to ten hours to pass with no eating, no going to the bathroom, just extreme focus.

There is another way to use your energy, especially if you are a generator and you manifest energy for other's use. You can start small by sharing your energy (money) by investing in companies you trust. You can work for others in a business that aligns with you. Most people work for a company, but does that company really align with your beliefs?

How is it that the evil got to where it did? *We offered our exchange of energy.* For example, when you pay to see a movie, when you exchange your energy for what you are watching, you are condoning everything you are watching; love stories, horror films, action films, every scene from those movies you condone and accept. *In the Ether, there is no make believe. If you feel that watching make-believe moves are someone's thoughts put into actions, and those thoughts are "not real," then what is prayer?* What are your goals? What is your love? All just make believe? No, your ideas, your thoughts, what type of energy you consume with your senses become your reality.

Unbalanced Focus- Unbalanced Energy Exchange- We are the sum of our choices

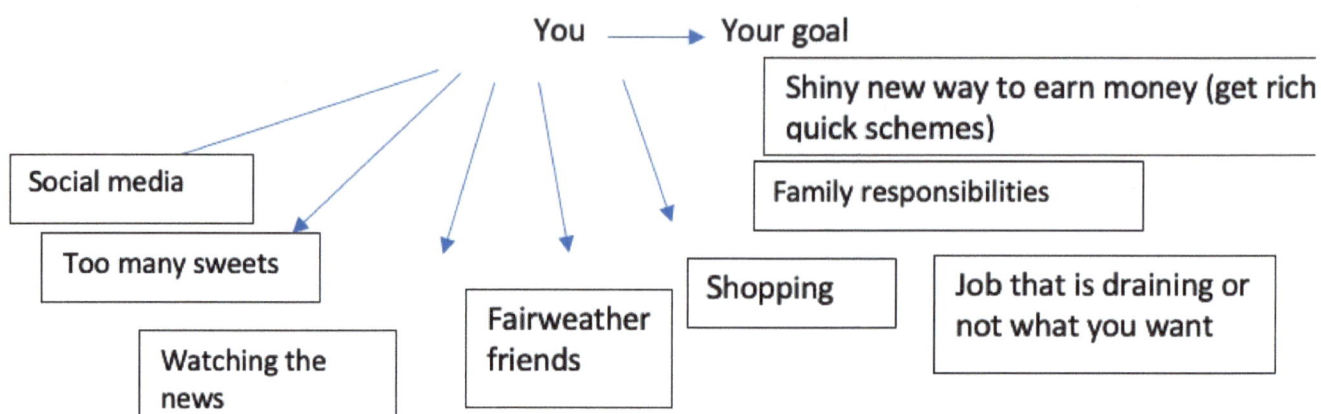

You ⟶ Your goal

Shiny new way to earn money (get rich quick schemes)

Family responsibilities

Social media

Too many sweets

Watching the news

Fairweather friends

Shopping

Job that is draining or not what you want

What are you focusing your energy on? Education to develop your skills? Relaxation? Relaxation is great, but how are you relaxing? By watching a horror movie? Social media? Eating sugar and drinking alcohol? What about listening to music? What kind of music?

Am I saying you have to change your life habits? Yes. It depends on how much energy you are wasting on thoughts and actions that are not serving you. The changes do not have to be difficult. We didn't get to a point of stagnation overnight. We are not who we are by one choice. It was a series of choices that got us to where we are now. Identifying energy leaks will take time and focus. This is a task that will never be 100 percent done. We are spirits on a human journey, constantly learning new things. So enjoy where you are, but if you are not happy now, identify why. Accept where you are, change what you need to, and move in the direction of your purpose.

Learning Discernment

What feels good? What feels bad? First off, we need to acknowledge our feelings. We have been taught to suppress how we feel. That still, small voice that says, "Hey, this isn't a good situation to be in." Santa's lap for example. Or following someone who is a hypocrite or, worse, a narcissist. Being still and trusting your gut takes time and practice. For most of us, it has been stripped from us from the time we were little. Being told to be affectionate with people we felt uncomfortable around, especially family; being told to sit still, comply with what an authority figure is telling you—it's all programming to strip away discernment for what makes you feel is okay and what is not okay.

Concept 5
Expansive Poetic Thoughts

Every soul is a drop of water in a constantly changing order in the cycle of life. One day you are a part of a cloud. Enough drops cling together, and it begins to rain. As you fall to the earth, some spatter on windows; some reform into puddles that are consumed by a bird; others are absorbed into the ground, used by plants, evaporated by the sun, or absorbed into the earth to become part of a stream, a river, a lake, the ocean; some become locked in a glacier for a million years, melt, evaporate, splitting into your smallest particle, dissolved—but not really.

Ever flowing, no real form, a part of everything, and never on your own. You are strong enough to split rocks, move continents if you band together with others. Weak in the light, you fall apart. You become a mist in the presence of the heat and light. In this state, you transform. In the clouds again, the electricity flows through you, hovering above the earth, stasis, cryogenics, observing what is below and above you. Is there oxygen and hydrogen elsewhere in the world?

If you ascend high enough, do you become part of it? Could you ride a lightning bolt in that state to another dimension? We are like water, changing states of matter, always the same components. Water is part of life on this earth no matter the form.

We are the water. We are the life in the light of creation. We hold within our bodies the primordial waters of creation, the waters that never left this earth.

Through water, vibration and frequency are possible. In the waters of creation, a baby is born. In waters, we transport our DNA as templates for reality. The original blueprint. Each droplet carries the secret to unlocking the mysteries. We must be present, whether in a cloud, a river, the ocean, a person. We create whatever we want through the vibration of the waters. The water is within us.

This is why humans have been separated, segregated. We were told one was different from another based on the shape or color of the vessel. Whatever the vessel, the water is the same water. The drops must be unified and resonating at the same frequency in order for the electricity to flow. The electricity/vibration creates light, the light illuminates our path so we can change, learn, grow, create, manifest. This is how the Law of Creation works.

Our bodies are mostly water. Nothing in this world is static. It is eroded, burned down, built up, and transformed based on interactions of the earth of what is within and what is outside of our planet. Our ideas are the catalyst, and it can be done with one single catastrophic event, where so much pressure is built up either in the earth, the sun, or human souls, and then a release, a title wave, a volcano, a pop inside our own bodies that aligns the energy so it can flow like water once again. Change can also be done with a single grain of sand in chaos.

Pure resonance: which is the soul fully incarnating in the body, firing on all circuits; where energy flows perfectly through the muscle, tissue, bone, tendon, kundalini, and the waters.

An Introduction to Manifesting:
The Law of Attraction and the Law of Action

Steps to Manifesting:

1. It all begins with an idea. An idea that comes from your mind, from the Ether. It can be a thought that drops into your head unexpectedly. A moment of inspiration while daydreaming or from an actual dream. Speaking is a way to bridge between the Ether and the grounding of the thought into the world. If you share an idea with another person, you have someone to bear witness to the thought that brings the energy into reality. There is a saying that the best place for ideas is a graveyard. It is full of ideas that had no action around them. So don't let your ideas and inspiration go by the wayside.

2. Next, in order to bring this idea into reality, you must speak about it *and* write it physically on paper or an organic substance. Personally, I do not think typing on a computer counts as it does not become physical until it is printed out. Technically, all electronics exist in the Ether because electronics are a system of energy existing outside the realm of physical touch. Yes, you can look at it, but you can't hold it in your hand beyond the conduit of an electronic device. Technology is also a closed circuit, not fully interacting with the Creator's Ether. The energy in this circuit is run through a program or programming. This is why when something is "written in stone," it is physically grounded into the Earth plane.

3. You must put actions behind the thought and written words. If there are never any actions taken, nothing will happen. You can pray about something all day. If you are thirsty and you pray for water, but never physically get up to get a cup and fill it with the rain that's falling outside (that the Universe has brought you because you need water), you will continue to be thirsty out of sheer laziness.

The Law of Attraction is really the Law of Resonance. This is why we say that something is resonating with us. It is the frequency/vibration that we find pleasing to us or repulses us. Do you value what I value? Do you believe what I believe? It is how we are attracting. Our mind can want $1,000,000, but if our mantra is "I hate money" or "Money is the root of all evil," you will never attract it. Why? Because your resonance, your frequency, is off. Your thoughts and your physical needs are in conflicting vibration with your words, and your vibration is actually repelling what you want. This is also part of being congruent. We will learn about this concept later.

How do we change our frequency? We begin by using our imagination. Exploring thoughts of what makes us happy. What is your desired outcome? Then keep track of your thoughts. Are all of your thoughts going in one direction? Anything we do first has to be held in our mind, in the Ether. Then we need to act. The book of James (Bible) basically says, yes, pray for what you want, think, meditate on what you want, but if you never act, you'll never get it.

How do you set yourself up for attracting what you want? Do you need more skills? How do you acquire them? Are you physically in the wrong place? Do you like where you work or where you live? Do you *love* it? Could you envision living somewhere else? If so, move. If you cannot move, then develop a skill that will get you the money/energy you need to move.

If you are not in the right place physically on Earth, you will have a difficult time manifesting enough energy to get you to where you want to go.

Are there occasions when we feel we have to stay where we don't want to? Yes, we put constraints on ourselves all the time. The only real reason to stay somewhere, and only for a short period of time, should be for love. Taking care of a parent or being close to your child. Maybe for a loved animal that needs to be on a property such as a horse that can't be moved. Other than that, head to where you want to be. A good tool to use is www.astro.com on the site click location astrology and astro travel.

If you thought about where you are drawn to and you really don't find an emotional charge to move, how is the energy in your space? Create a space just for you and your thoughts. It can be a chair or a space on the floor. Some cultures create a room with an altar. You need a space where you can clearly think and listen to your thoughts. That is all that is necessary. Now if you want to go all out—candles, crystals, motivational pictures—go for it. Once you are in the right space, it's time to work on your thoughts, energy, and frequency.

What thoughts are coming into your mind? If you find the thought important, write it down. Keep a journal of your daily thoughts and your dreams. Begin to ask yourself questions. This is the theory of going within. Act with faith but without definite expectation. Allow the Universe to rise and meet you. This is the part people call the Law of Attraction or Resonance. Act on the thoughts you have during these sessions. Practice making choices that will lead you in the direction of what you want. If you make a choice and it doesn't seem to work out, make another choice. The Universe will begin to bring you the information you are seeking. You will begin to be suggested videos or books that will help educate you. You will begin to meet people that will help bring you to the next level.

Let Creator/God/Universe work to get what you can't. Be open when new people come into your life. Accept their gifts or their wanting to work with you. Allow others to support you through their thoughts and actions. Watch as your congruent actions and your acceptance of help from others begins to open gifts the Universe is bringing to you. Watch as your bank account grows from you being more successful. Pay attention as you become happier, healthier, and sleep more soundly as you come into alignment with your purpose, naturally, almost too perfectly. As you're making choices and going down the right path, you will start to see synchronicities.

Synchronicities are where you will begin to see repeating numbers, a name appears repeatedly, or you could run into the same person in several locations, all of this seeming random at first. You will start to notice the repeating nature around you. You begin to notice "signs" to visit a certain location, maybe a vacation, or maybe a place to meet a new person in your life. You see signs about what to do for work or skills to learn. If you're musically inclined, certain songs will play on the radio that will deliver you a message coded only for you.

Start learning about what you are truly interested in. I have a sneaking suspicion it is not something you learned in a government school. A skill should be something that leads you closer to your passion. If you are interested in painting, astrology, or electrical engineering, learn it. If you need to continue to work, which we need to do in this world to earn money to live, wouldn't you rather work in a place that makes you happy? I've seen people continue the daily grind because "the money is good," but they are totally unfulfilled to the point of exhausting their body and their mind. The money may be good, but there is no real life progress. Meaning you've climbed the corporate ladder, for what? To be burnt out and unfulfilled? There are those that think they can make it all on their own.

On the other hand you may absolutely love what you're doing. You may be a health-care practitioner or a massage therapist, but how many people can you actually see in a day? Do you have anyone to partner with or to help you gain new clients if you are slow? If you don't let anyone else in to share ideas with, going at life alone can be, well, lonely.

Your frequency goes out into the ether and begins to pull toward you those that resonate with you, that have similar ideas, business partners, love partners, your "tribe." It can also help manifest where you want to live, how you want to design your life.

Concept 7
Structure of Energy Interaction: Love, Gifts, Work, Passion, Skills, Thoughts, and Actions

Love—Love should be innate to every human being: the giving freely of energy without the expectation of anything in return. However, because there must be an equivalent exchange of energy, the love you give would have energy returned to you in an equivalent amount, as in the example the love of a child or the love of your life partner.

From Etherical to physical. The thought, the emotion behind the action from two human beings together—a love, passion, desire—created a spark from the Ether. Through love, the energy is carried on, nurtured and manifested into the physical.

Gifts—Gifts are skills that come naturally: listening, compassion, wanting to take care of others, calming others, empathy, strength, emotional knowledge and understanding, problem- solving. We can deliver our gifts through our work. Our gifts are transmuted, converted through work into an energy that can be accepted by others for payment. Our gifts that are given freely to others still has an energy exchange and is emotional and no less important.

Giving gifts that mean nothing is lost energy and actually has a negative impact on the receiver. Donating without emotional attachment can be beneficial. If you donate out of sadness, anger, hate, those emotions can linger with an object. If you don't believe any of this, then you have been wonderful stewards of your energy. But have you ever seen a scary movie or been told a story where an object retains bad juju. (now that we know about preserving our energy we will set our boundaries appropriately with movies LOL) Okay, maybe it's not that extreme. Have you ever received a gift from someone, say an ugly jacket, totally not your style, and in the wrong size? Maybe the obligatory gift of a candle with a funky scent? How did it make you feel? You probably felt like the person doesn't know you at all. "Gifts" like that take away energy and cause the person receiving them to lower their vibration. So you may as well not give anything if you're that type of gift giver. Gifts in the joyful way of giving and receiving complete a positive interaction where both parties involved feel fulfilled and a circuit of understanding and love is complete.

Work—What are we put here to accomplish? What do we want to experience in this lifetime? What do we want to feel? Our work is tied to what we want to accomplish in life and leave behind as a sign of our being in existence here. As kids, we only want to do what is fun and exciting. We want to be with our friends, others who are similar to us but different enough so we don't get bored. This is what we should look for in our daily work. What we give our energy to in exchange for other energy (money).

We work our way through energy exchanges from the moment of conception. The moment energy is exchanged and a spark is generated, a gift is given and received. From Etherical to physical. The thought, the emotion behind the action from two human beings together—a love, passion, desire—created a spark from the Ether. Through love, the energy is carried on, nurtured and manifested into the physical.

Touch—Gifts and work can be implemented through touch. Usually, the first thought is a tactile job (massage therapist, mechanic, hair stylist) but can also be medical such as nursing, veterinarians, and the physical interactions of how we treat beings and objects in the physical world, how you play with your children (board games, reading, rough housing, being outdoors). The touch of your energy transfers through the people, animals, and objects and create an exchange of emotions and memories. What do you feel while you're looking at your Grandpa's empty chair after he's gone? How many good movies did you watch together from that chair as a kid? What about an old fishing pole of your Dad's, that stuffed animal you played with when you were in kindergarten, the book you and Mom read every night for a year?

Touch can even be a location. For many, it is the feel of sand between your toes, the smell of the salty air and the sea breeze on your face, the crunch of leaves, the campfire smoke during fall, or the snow and pine scent at winter. I am mentioning scent for a reason. Technically, tiny particles are interacting with your body inside your nose so you can recognize your surroundings. This is a physical interaction with the physical world. The Etheric interaction is how the scent makes us feel and the ability to instantly recall where we were, what we felt, who we were with, the emotions—everything, all in a second.

The residual energy of touch is powerful. It is how the energy transfers from the Ether and manifests physically with the strongest potency.

Passion—This is doing work you enjoy and could do this work every day for hours on end and not be tired. This is your life's work, your purpose in life.

Purpose/Passion—What would you do if money did not exist? What would you do every day if you had everything taken care of, food, shelter, love? Who would you want to help?

If you are not passionate and personally and soulfully invested in the work you are doing, if you are not serving others while gaining energy for yourself, you will be drained, and experiencing life to the fullest will always be out of reach.

Skills—Skills can be natural born gifts or can be developed over a lifetime. Skills are what is used to create an exchange of energy where the energy received is monetary or gaining in physical wealth in some way.

Thoughts—Ideas, concepts, downloads from the Ether, matching others who you think have beneficial skill sets to form a group either for social or work function, how to accomplish a task. You could be problem-solving, but it doesn't have to be a problem, it can be how to more efficiently complete a task or a type of mastermind group of learning by sharing thoughts with one another. Mathematical equations, quantum theories, pushing the limits of what is possible all come from downloads from the Ether about what someone believes could be true.

Spirituality—Is the understanding that a power outside of yourself exists and that you are a part of it.

Energy Exchange: Know your worth and what value you bring to this lifetime.

Wu World (Ether/Spiritual)	Reality World
Gift	Skills
Thoughts/Download/Channeling	Education
Touch	Interaction with people, mechanical, building
Love	The energy it takes to care about the people around you or the positive energy it takes to produce a product or service
Work	What you get paid to do, you can also barter for products and services
Passion	Purpose in life
Spirituality	Quantum entanglement

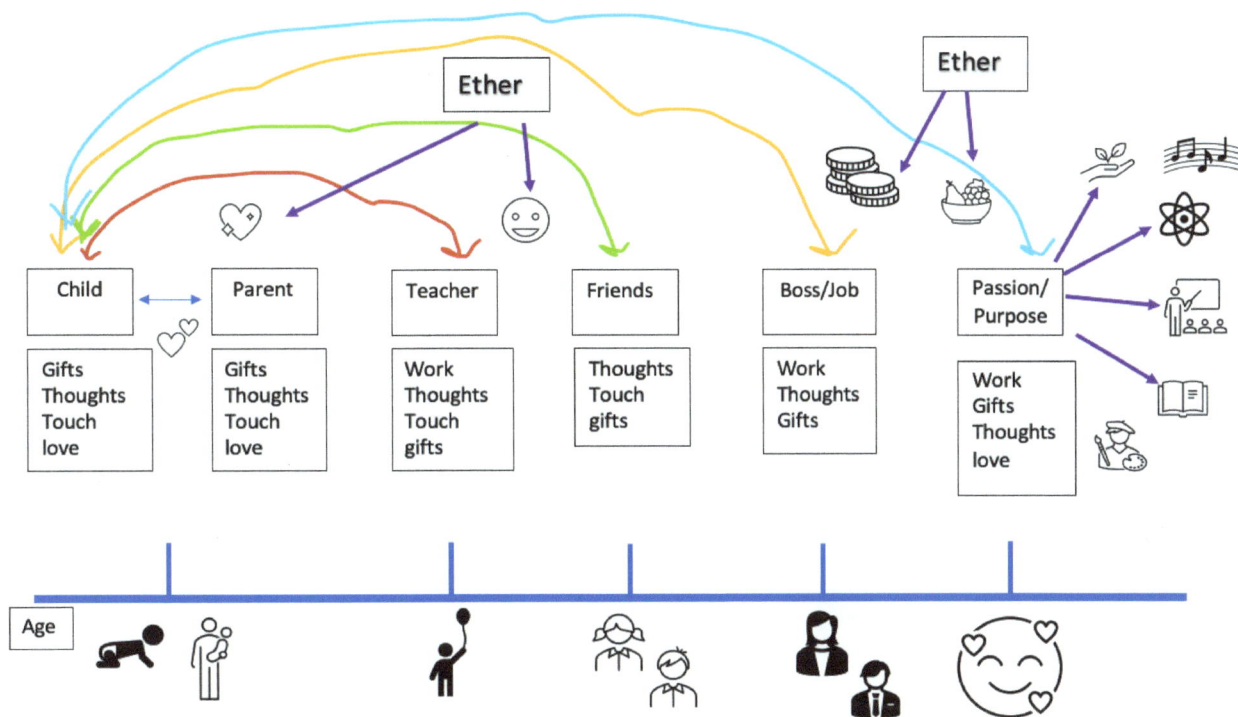

When we think of being successful, we think of sharing our gifts with the world. In return, we will receive energy back in the form of gifts or money. This is the first moment a human being is pulled from the Ether, the spiritual, into the physical. The child then continues to receive energy from the Ether in the form of love and emotional support from their parents. We all, throughout our lives, remain connected to the Ether. Love, emotion, the ideas of spirituality/religion keep us connected to the layer just outside of our physical reach but is very much a part of our ability to remain functional in our bodies.

Most of us can understand the love of a parent and their child or the interaction of a caring teacher and their student. Friends are our first step into the realm of peer/social interaction, the first form of relationships. It is a mix of giving and receiving gifts as well as our first step into bargaining for energy. If you let me play with your dinosaur, I'll let you play with my remote-controlled car. If you push me on the swing first, I'll push you next. Let the learning from life begin.

We receive thoughts and gifts from our teachers and friends. We experience this physical manifestation of emotion through touch. Business is where we have been taught to silo ourselves. We begin to compartmentalize ourselves and pull back on love and touch to keep it "just business." That is what we've been programmed and trained to do. The energy exchange is for only a part of ourselves. We are only offering a part of ourselves to this job in exchange for the money we need to live. This creates the fractaling of our physical and emotional self, and it is all by design. We are made to split ourselves, business and personal, to keep from presenting our whole authentic selves to any one person and having a thoroughly 100 percent interaction. It prevents us from being fully present, fully honest, and from being in our passion, thus preventing us from ever experiencing life to the fullest.

Do you know one successful person, one wealthy person who doesn't pour 100 percent of themselves into their life's work? Do you know an abundant person who isn't passionate about their purpose, "living their dream" or "living their best life?" Do you think they're just living their best life from 5:01 p.m. to 6:00 a.m. when they get up to start getting ready for work? Or is their work their life's work their passion?

Don't you want to be living an abundant life 24/7? That's not saying you're not completing your life's work, but your idea of work needs to change. What impact are you most wanting to create? What do you find purpose in? If you find that, it's not "work," it becomes a free flow of positive energy exchange between all parties.

The understanding of all of energy exchange is actually a mathematical equation. Both sides must be equal. The simplicity is, there is no positive or negative in this universal equation, only a total amount of energy units given. I don't have a specified quantifiable unit of measurement of energy, nor do I even know if humans have a way to measure Etherical energy or not, only that each of us knows the difference between being physically drained, emotionally drained, or physical and/or emotionally full. Our human energy is comprised of both in this Universal Law of Etheric Energy Exchange on earth.

We will start with the work relationship since this seems to be the number one driver of anxiety and fear and how we obtain our energy (money) to live the life we desire.

You give of your skills, gifts, and thoughts to your work, which gives you money. But money is not the only energy you receive from your job.

When you leave your job, how do you feel? Stressed, undervalued, angry, tired, agitated? Is it effecting your weight, stress eating, or not being able to eat? Does your family notice a difference in you while you're either preparing to go to work or when you get home from work?

Say for your job, you are hired at a local restaurant. You are making $15 an hour.

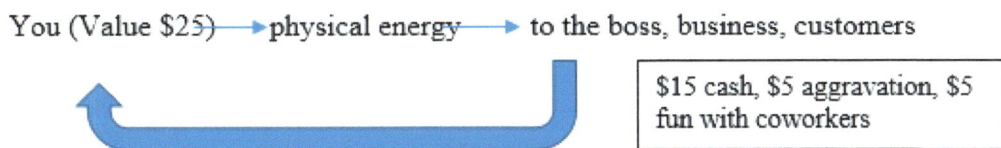

You (Value $25) → physical energy → to the boss, business, customers

$15 cash, $5 aggravation, $5 fun with coworkers

If you feel like your worth is $25 an hour, you are receiving back the $25 in energy, $15 of it in physical money and $10 energy in aggravation, anger, depletion. It can also be the opposite. You can feel you're worth $25, but receive $15 in physical money and $10 in positive energy because you're working a job with your friends or learning a skill you find interesting. Money was created as a way for us to transfer to someone else who has what we need but does not need or desire our specific skill, gift, or energy at that moment. For example, you go to work at a manufacturing plant making nuts and bolts. The owner of the company needs your energy to help make the nuts and bolts. You personally do not need these nuts and bolts, but you give your energy in exchange for money.

Money complicates the system of natural energy exchange by adding a layer not only on top of the original bartering system, but also a layer to the Etheric energy pull as well. The bartering system resurfaces when economies go soft; when countries, kingdoms, or empires fall; or systems become obsolete. Why? Because the man-made infrastructure has crumbled back to a natural state, dissolving the additional layer of having to use an external agreed-upon system.

Block chain technology and cryptocurrency is attempting to merge the Ether and the physical by creating a way to exchange on the basis of bartering, but without having the oversight of an external system outside of the two parties in the energy exchange.

Now we understand that energy exchange units are equal no matter if it's positive or negative, physical or emotional, the exchange gets filled, and the circuit of energy exchange is complete.

Concept 8
Own Your Energy

Types of Actions

There are actions that give you energy and actions that take away your energy. Actions that give you energy should involve the gifting of your time, skills, and emotions. Gifts are usually in the form of preparing a meal for your family, raising your children, changing a diaper (maybe), or volunteering with an organization you value. We are compensated energetically through emotions, usually love, a sense of accomplishment or gratification.

Actions that take away your energy is called work. This is why we are normally compensated in a physical form for our work, either with money or other material products.

This concept sounds simple. When we were kids, adults would ask us, "What do you want to be when you grow up?" Not "What makes you happy?" not "What do you enjoy doing?" not "What do you find interesting?" Because as a child, we're not going to answer with a "job that pays us well." We're not going to answer with, "Whatever's convenient at the time." Why? Because we don't know any better, but also because our mind and decision-making is very linear and very clear. We go with our intuition. If we find a subject interesting, we'll pay attention; if not, we daydream. At an early age, the programming starts. How to act, how to feel, what is expected of us is engrained in our thoughts and actions. Outside influences and external propaganda bombards our thoughts and emotions and influence how we feel about everything, from what we wear to how we feel about ourselves, how we should react to topics of the day, and what we should do with our lives. Its never about what *we* want to do with our lives or understanding how we feel without outside influence. All the external noise influences and messes with our intuition.

We will begin to explore how to shift our mind and observe the energy exchanges going on around us, influencing us. Why do we feel so drained? How do we make better choices and stop repeating scenarios that only harm us and keep us on a never-ending loop of two steps forward one step back, all while never realizing we're just walking in a circle.

Energy Exchange of Feelings

What you put your money toward is what you endorse. For example, if you pay to watch a movie and that movie contains violence, abuse, and/or horror, *the Universe does not distinguish between humans' reality and*

38

fantasy. Universal Law will consider this as an energy exchange of endorsement; thus, you endorse what you exchanged money/energy for.

The same goes for money for an uplifting movie, supporting the underdog, romance, comedy. Money/energy is exchanged for education, food, housing. Money/energy is exchanged as a gift. So think consciously about what you want the Universe/Creator/Spirit to think you're endorsing. The process of this thinking happens through a series of energy signatures, attached exchanges of energy, a series of equations that happens trillions of times a second on this planet.

Everything in this world has an energy signature, an imprint of a charge of creation. Objects by themselves are not inherently good or bad; the energy placed on them by a being of creation is what creates the charge. This begins the building blocks of any equation. This positive and negative energy exchange is different from the Universal Energy Exchange where the charge of the units do not matter. Any exchange of energy carries a charge. There are physical objects present in the physical world; there are ideas contained in the Ether. Ideas/thoughts are energy that has not yet manifested into the physical world yet can still influence it. Have you ever stepped into a room where two people greatly disliked each other? You can feel the charge in the room, the pressing negativity. The same perception of emotion can be felt at a gathering or party where people are laughing and enjoying each other's company. You can see happiness in a smile or hear laughter, but you can't touch it. Happiness has not physically manifested as something you can touch, yet a feeling of euphoria and contentment has passed through everyone in attendance. The charge has passed through each electrical body in the room, including pets. This is part of the quantum exchange of energy between particles.

You don't want to receive "dirty money." Many people are familiar with this phrase. It usually has to do with money received from an illegal activity and the cops have probably marked the bills for tracking. So this is definitely money you don't want. Let's look a little closer. There is negative energy attached to the physical bills. Law enforcement has to physically identify the energy signature of the money in order to track the money so it cannot be used and put back out into the world for public for use. Some people, even if the physical bills weren't marked, if they knew it was "dirty money," they wouldn't take it for fear of bad karma. So even in modern society, there is still a spiritual energy or negative energy aspect to our transactions.

The unfortunate thing is, we participate in "dirty money" exchanges all the time without even realizing it. When I say that everything has an energy signature, that includes money/energy that you have earned. The person or business paying you the money may not be negative, but if you hate your job, or even just feel like it's not worth it, you have tinged the transaction with a negative energy. Anything you purchase with the money you earned now has a negative charge to it. Food, clothes, your living space will also carry a low vibration or negative energy.

I know most of us have experienced less than desirable jobs, jobs we've taken to make ends meet, had crappy bosses, or even abusive work environments just so we can have food and a place to live. I get it. However, this is the cycle. The two steps forward, one step back. The merry-go-round of life at the horror-show carnival. Receiving money in this way, by attaching negative energy to it, even though it's subconscious, will break you. It will cause depression, anger, sadness, illness, and leave you no time for yourself or your family.

By changing this singular charge of energy that you're attaching to the money you're receiving, you will begin to see a change in the rest of the energy exchange equations in your life. Right now your job could be crappy, your dead-end job, the place you cannot see a light out of right now at this moment; but know there is a way out. And this is the start. Even if you do not believe in a Source, a Creator, God, say out loud as you receive your paycheck, tip, income, or gift, "I appreciate and joyfully receive and accept this energy." And freakin' believe it! Don't half ass this part or you'll never get off that damn horse on the merry go round at the carnival from hell.

Now that we know our money carries an energy charge, how does this all really work? Being clear in your thoughts and clear in your intention is what attracts your idea from the Ether into physical realm. In Hinduism, this is called a mantra. Mantra literally means to physically manifest using sound. In the US, a mantra is something we repeat to make ourselves feel better, to keep focused on our task or goal. When we say, "I am not a money person. I'm not greedy" or "Money is the root of all evil" or "Rich people take advantage of the little guy," you will never have money. You will never be rich or abundant. You have verbally blocked the energy flow of receiving the frequency, the resonance, the energy of currency. While we never want "dirty money," we certainly can put out a clear message of receiving gifts, being compensated for our work, and receiving positive energy and abundance into our lives.

We receive positive energy from our work or gifts. When we are born, we are gifted life. We did not do anything to receive the energy. The Universe united material from two different beings and formed a new spark. Then your parents gifted you love. They do not expect anything in return. Their love converts energy into you. The transition from the Ether to the physical has a few converters that alchemize the energy or transmute the vibration to allow the manifestation to happen. Love is one; actions are another. Your parents' energy is converted into clothing, food, shelter. It is this transfiguration of energy from the Ether that sustains life. Abundance also does not need to involve physical money at all. Houses, vehicles, food, clothes, vacations can all be gifted from the Universe at any time, as long as the channel is clear, and you are open to receiving. Food and clothes are the most believable so start with manifesting these gifts. A close parking spot at the grocery store, hitting all the green lights at the way to work are also good to practice manifesting.

Concept 9
Prana/Chi and the Etheric Energy Body

Our bodies have an energy system that courses through it. We are not just bones and muscles, through out our nervous system, electric energy flows. Much like our computers have hardware and software, our bodies also must have an electric current to flow through it, or it doesn't work.

The magnetic field that surrounds us and the Ether supports each individual being's body. The energy itself within each body functions based on that magnetism and the spark of life that we each have. Generated by our hearts and flowing through our body systems, we have two circuitry boards. One is physical, which runs along and through our nervous system. This is called the *meridian system* and what acupuncturists use to treat the body. The second is the life force that runs through us like a DNA helix structure, the *kundalini.*

This energy—which in India is called Prana, the Chinese call it Chi, some call it Life Force, Jedis call it The Force. This is what powers our body. This is what our soul uses to stay connected to the hardware of the earth. The organic technology that is our body is separated into parts.

The parts of your body that are energy centers are called *dantiens,* of which there is an upper, middle, and lower. The dantiens are segments of the body that help overlay our electrical system. Usually activated by breathing and body movements, this engages your body into connecting with your energy circuitry.

The energy passes through the Chakra system of which there are seven concentrations or energy vortexes. Some may say more, but these are the basic seven: root, sacral, solar plexus, heart, throat, third eye (mind), crown (soul).

Chakra means "wheel" in Sanskrit. The swirling vortices of energy passing through our bodies help to fully integrate the soul with the body. The lower the chakra, the lower the vibration. Our primal instincts act out of our *root chakra*, near our reproductive organs and at the base of our body. It is also important to be in touch with the Earth through our root. This is how earthly energy flows up and into our bodies. If this root becomes damaged physically or emotionally, violated physically, or blocked because of the mind either intentionally, unintentionally, or prior experience, the energy will stop resonating. Your body will stop resonating with the Earth, and sickness will set into the body. Exploring this part of your body is the basis for your existence on this Earth.

Sacral is where the energy of creation starts. This is how your ideas from your soul and mind will flow down and out through this circuit or passageway. If your ideas are not fully formed, you will never birth them into

41

creation. The sacral will resonate if you are attracted to another person. Bringing energy through your root into your sacral is the first step to bringing energy, especially Earth energy into your body.

The *solar plexus* is the identity of self—who we are in this lifetime, who we choose to be, our confidence or anxiety. It is the butterflies in our stomach, the speaking from the core of our being.

Our *heart chakra* is green in color, healthy and growing like the plants around us. It is our love spreading to others through the vibration of the earth and through the Etheric field around us. It is no coincidence there is so much green on this Earth.

The *throat chakra* is blue. It is how we communicate our message to speak to the world, to lift our voices to the sky and resonate. Again, there is no mistake that the sky is blue.

Indigo of the *third eye*. Purple is the color of royalty. Why? Royalty is considered to be empowered and enlightened, the ability to transcend the Earthly thoughts and be a spiritual being. Violet is the connection with spirit, the Ether, and the Akashic records. Being able to pull energy into yourself, your third eye, your mind and into your brain, allows the electricity to flow through your body to all parts of yourself. For the electric spark of the idea to enter your mind/your brain, speak through your throat, feel it in your heart, your gut feeling, and then birth into creation. It may seem odd for a man to birth, but there is the mechanism of birth, and what you have also has to flow out into creation to become physical.

There are other levels of energy above and below our bodies. If you are interested, start researching. Learn for yourself how the energetic mechanisms of your body work.

One of the reasons Druids had no written history is because they did not need it. Everything is held in the Ether, which was accessed through the same electric field and integration of energy, soul and body. Ideas are held in the Ether. What you think is your own idea is actually born from the collective energy of those around you. This is why when a new invention comes into being, it can happen from several people across the world. Books with the same subjects are published around the same time.

Chi is also separated into systems called meridians of which twelve organs are used as circuits to pass electricity through. Chi is broken into masculine (yin, right side)—heart, spleen, lungs, kidneys, liver, pancreas—and feminine (yang, left side)—small intestines, stomach, large intestine, urinary bladder, gallbladder, and the pericardium.

If you look at these systems, you will see it looks like circuitry. We are a circuit plugged into the motherboard, that is Earth. If you break these systems down further, into acupuncture points, you can begin to see how to start turning on your energy. There are some of us who believe our systems of integrating mind, body, and soul have been targeted, damaged, programmed, or other some such degradation for our lifetimes and possibly for thousands of years.

One thing is for certain, going back to the water principle, the Ether holds the templates. The water facilitates the holding of those templates, and once the Etheric energy passes through, the body, our circuity and electric field can be corrected and return to its original template, or original design.

It is no different than eradicating a virus on a computer system. With an antivirus program, a computer can be corrected by eliminating the bad program and then reinstalling the correct system from source.

Our mind perceives us as solid holographics through our organic software system. Feathers are light and soft and elephants have rough skin and are heavy because that is what the mind is programmed to believe or programmed to interpret. Reality is however you are experiencing life. It is what is real to you, but it is also what is accepted by the collective thoughts of others in this realm.

Have you ever discovered an untruth? One that you were programmed so hard into believing that you swear it was the truth, until it wasn't. Most of the time our understandings have to do with the human body. The best example is, for years it was thought that the human body could not run a mile in under four minutes. Then when Roger Bannister broke the record, others were suddenly able to do it to, and the belief in the "truth" was no longer so. Humans think this now of childbirth being painful. It is so hardwired into the construct of humanity that the idea became "remember" how childbirth is supposed to be. An apple does not grow a seed of pain, ladies, but the thoughts, the *beliefs* of billions of humans does imprint on our reality. Another process is scarring, regrowth of skin, bones, and cartilage. If given the right conditions, surgeries can be done without scarring. Skin can regrow after a natural removal of unhealthy cells. The problem is, humans do not understand how to create the environment for a positive outcome; so therefore, it is impossible, false, a lie, until it isn't.

I am saying this because if you know the resonate frequency of a healthy liver and you play that frequency through the scalar field, the field knows what a healthy liver looks like, and it will replay those frequencies into a damaged liver. The cells impacted by that frequency will change to fill the void of the construct. The construct, the fabric that this reality is made up of is programmed in the way the structures fit together. Meaning, humans have two arms, two legs, and a head; mammals also have four arms/legs and a head; birds, the same. Each species has a predetermined construct. A template. A blueprint. A design. This is why the energy field works, how the vibration can change and morph to fill the gaps or repair the breakage or damage and become whole.

Taking care of our energy body is important to the health of the physical body. Eating clean food, drinking clean water, working out your muscles, having healthy structure supports the circuitry in our body. Being outside in nature and around positive people, you will be able to support the electrical flow of your body.

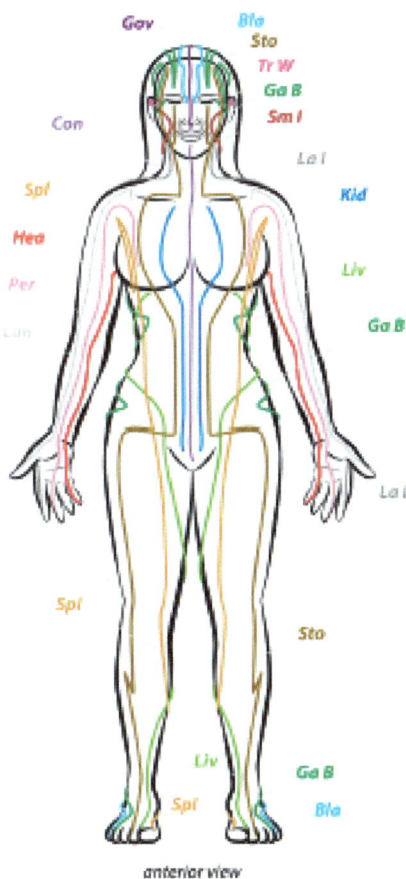

The Body Meridians

Two Centerline Meridians:

Conception Vessel
Governing Vessel

Twelve Principal Meridians:

Stomach Meridian
Spleen Meridian

Small Intestine Meridian
Heart Meridian

Bladder Meridian
Kidney Meridian

Pericardium Meridian
Triple Warmer Meridian

Gall Bladder Meridian
Liver Meridian

Large Intestine Meridian

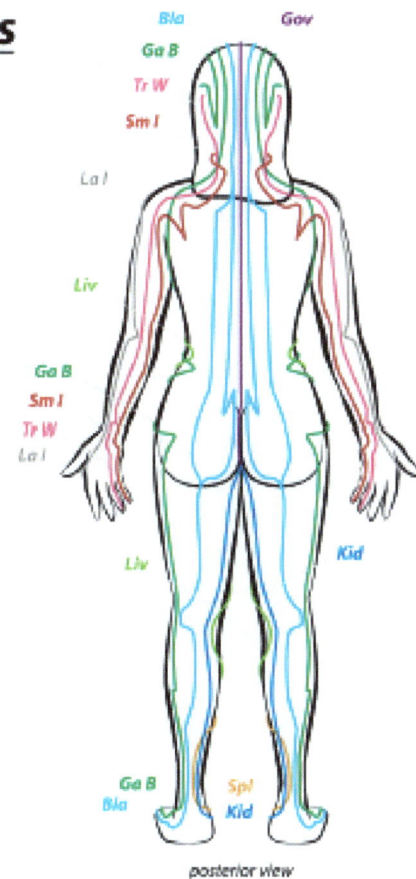

anterior view

posterior view

Concept 10
Activation of Kundalini

Kundalini is the energy system attaching our body to the Earth. It operates through our body and interacts with the spiritual realm. This energy system supports your discernment and your ability to protect yourself from outside influences.

This is why sex was made dirty by some religious dogmas. If humans knew we had the ability all along to communicate directly with God, without an intermediary, the religious institutions would not be able to hold their position of power in the world. This is why the demonic attacks children at their root (pedophilia). It prevents the body from manifesting our toroidal field and connection with the earth and the Ether. The caduceus in the medical field is a symbol of our DNA and also of how the Ether flows through our bodies. The magnetic field surrounds the body, and chakra energy flows through the physical body. Kundalini carries our Etheric bodies.

Kundalini etheric energy body

Magnetic field physical body interaction the interaction between the physical and etheric

44

Activating the kundalini takes an activation of the chakra system with a gland in the brain called the pineal gland. We have water in our pineal gland, which is encased in a crystalline structure. Calcification of the pineal gland prevents Etheric energy from permeating into our physical brain, severing our true connection to Source/God/our higher selves. There are many accounts of humans "glowing" and thus being referred to as decedents of the God(s). Hercules, Noah, Jesus—all glowed because their pineal gland radiated light through the sound of their voice and the water within their bodies, focusing energy through the pineal gland to manifest or have visions through the third eye. Essentially, our vibration resonance pulls energy from the Ether, through the water in our body, through our organs, and out through the actions of our body.

There are many magnificent old-world buildings used by religious people to gather energy in one place. However, the building isn't the church; the people are the church. Human beings are the embodiment of the Christ and the Christ Light. Our body uses water to resonate when we speak and to bring forth life through our words. We speak our ideas into creation first, then act upon those ideas. Thoughts become words, words become actions, actions create the world, etc. etc.

We have been poisoned by an opposing force that does not want humanity to flourish and reach our full potential. Fluoride has been placed into many cities' water supplies in order to calcify the pineal gland. I live near a lake where the lake water is "approved" to be fluorinated. Approved by whom? Why? Why do we have so many additives in our food? What are the additives doing? Clogging our organs with toxins instead of resonating frequencies of creation. Why do we watch TV programs that program our minds into distraction? All of the matrix is a distraction away from connection to each other and to creation.

Activating our kundalini includes the following:

1. Clearing our organs of heavy metals (metals disrupt electric flow).

2. Clearing our minds of distractions. Discernment, ideas, inspiration cannot be pulled from the Ether if we cannot dream, meditate, pray. If the signal is scrambled from source on the way in because we're watching too many mind-numbing videos, we will never receive the signals we're meant to live our lives to the fullest.

3. Cleansing our pineal gland of corrosion. We must have clean water, not treated water, bottled water, fancy water. But clean, filtered through the earth, structured by the energy of the earth water. Did you know there is a town in Russia close to Chernobyl that was not affected by the radiation? They drank water that came from a stream that had series of waterfalls. The water they drank had a specific charge and structure that purified their cells, and the people excreted the toxins from their bodies.

4. Choose real food to eat. Choose what is minimally cooked, has one ingredient, or very few pronounceable ingredients. Eat as closely to the Earth and fresh as possible.

5. Weeds are medicine. The sterilization of the lawns around our homes is part of the programming. Who has the greenest, most pristine grass? Who cares? Who programmed us to think this way? What flowers are growing in your yard to feed the bees? What living ecosystem do you have around your house? Yes, this includes bugs, spiders, wildlife, birds, rabbits, squirrels, deer. Or do you live in a sterile environment? Heaven forbid you have weeds! But what exactly are weeds? Most weeds are medicinal or a food source high in vitamins and minerals. So why were you programmed to kill all medicinal plants? So you won't be able to care for yourself and you'll be dependent upon a system of managing your sickness with other man-made poisons. You will kill all the food around you so you can be dependent on "food" packaged in boxes containing God knows what at the grocery store.

6. Crystals, tarot, astrology, pendulums, runes, organic law, universal law: what stories have you heard about the previous words? From whom? Why? These are all tools to connect your spirit to the Etheric/spiritual realm, again, without a guard, an interpreter, or an intermediary. Why were these items or sciences forbidden? Again, power and control. Everything in life has a duality. Boiling water can cook your food, or it can scald your body. A knife can be a tool or a weapon, so can your thoughts. Think truly about why certain items or ways of thinking are banned. Who does it really protect, and who does it keep in a cage?

7. You cannot worship your sexuality as your spirituality. Kundalini is creation energy, which is why it is most often associated with sex, intercourse, and the tantric. Sexuality is important. Sex is important. Feeling creation energy by yourself or with a partner is important. What is most important is how you harness this energy. Harness it? In what container, for what? Well, in your body. Kundalini is used to pull energy from your root, from the Earth, up through your body to your third eye/pineal gland, and out into the Ether. Would you be surprised to know this is how a flower grows? Crops, trees—all plants grow through kundalini (kinda takes the sexy out of it, but...). The flower is growing, birthed from the Earth. It is fertilized by interaction with other plants and draws in water from the earth and light from the sun. Through the kundalini energy, all creation on Earth grows.

You harness this energy in your heart chakra and through your breath. After the energy has coursed its way through your body, this is the time to meditate, sleep (ask for a prophetic dream), create your art or write.

Think of who has harnessed your kundalini in the past? What programming were you watching, what movie, what club did you go to and flaunt your sexuality? There is a reason movies oversexualize their story lines. If you are not harnessing your creation energy, *who is*?

How do you pull energy from the Ether to manifest light? Using kundalini energy and through sound and water. This is why we make noises during sex, the vibration of the sound, the water of our body, and the water in the pineal gland help to produce the orgasmic, or Etheric, experience.

The old-world church buildings mimic this body of creation. Um, what? Yes. Churches originally were not places of worship but were built as energy centers. The towers contained large crystals, not bells. Every church has a basement that was either filled with water or other conductive liquid. There was a pyramid in Mexico that was sitting on top of a pool filled with liquid mercury. The building is like our body, using water, crystal (pineal gland), and metals (specifically copper)—all amplifying and recreating what is in the human body on a larger scale. Specifically copper wire was wound around a crystal where the bell in the towers are today. The wire was connected to the fluid in the bottom of the building, thus creating a charge. Where is copper concentrated in the human body? The brain, kidneys, skeletal muscles, heart, and liver. And what do people play at church? An organ. Humm, starting to see the picture?

Look up "Church of Jesus Christ Scientist" and look at the images. Somehow this church has acquired or "built" a lot of these "churches."

The copper wires run like DNA, carrying the frequency. DNA is carrying the programming frequency of creation, the template needed for the design of life. Why are there "organs" in a church that resonate and echo through the building? What used to be in the towers?

In India, you can see many temples, also with lingams, crystals, stones, and mercury that were used in conjunction with water to make what?

So "religious buildings" are resonating from sound. Light comes from sound. By understanding true science, creation science, we have a better understanding of how our world was created. Sound/vibrations make shapes. What is the sound of a healthy cell? What is the sound of a crystalline cell? Why are we carbon-based forms and not silicon or crystalline? AI is based on silicon. Certain materials cannot go beyond a specific frequency/vibration.

I can tell you when a TV is on in another room. I can feel it. Most electronics do not work well around me. I'm surprised I can use a computer and cell phone. My experience is, especially if I am in a bad mood, the electronics will not work at all. The body is intended to do everything electronics are being made for—instant communication, transportation. Once our body reaches a certain frequency, technology as we know it becomes obsolete.

All of this to say, kundalini energy is the energy of creation. It is housed perfectly in the human form. No matter how humans try to create this energy outside of ourselves, it will never be a replacement for what humans can create from within themselves.

Concept 11
Alchemy

The mainstream definition of alchemy is the transformation of matter. The Etheric element is the ability to change the frequency of thought. Love is a form of alchemy. Understanding is a form of alchemy. Forgiveness is a form of alchemy. By owning your free will and taking back the power of your choice, you will begin to draw Etheric Energy back into your vessel/body.

Say someone hurts you emotionally in a relationship. Out of the blue one day, they say you can't be trusted. They question your whereabouts and accuse you of cheating. You can remain hurt and walk away from this person. You can call them names and question why you were even with this person in the first place. Or you can choose to try and understand why your partner suddenly changed their behavior. This means you need to pause, reflect, and go within. (1) How is it that my partner's emotions toward me changed? (2) Is there anything in their past that would cause them to be insecure? (3) Did *I* behave in a manner that raised concern?

This process of *questioning yourself* first about the *situation* to understand all perspectives, questioning yourself about *your own behavior*, and then, lastly, asking for input from a *trusted external source* that is unbiased to the interaction is part of the alchemy. (4) The last part is to listen to the responses of the other party and see where the emotional pieces fit. Usually, it is not one sided. Your partner could have wounds from a past relationship that surfaces from time to time; but you, on the other hand, may have gone out either alone or with a person they didn't particularly like and not told them about it, thus causing the old wound to open and negative energy to pour out.

It's common to want to push all the "blame" on one person. In actuality, there is no blame. There is no right or wrong, only the communication or miscommunication of energy between two people. There are boundaries that need to be communicated, what you will and what you won't accept from a partner. If we keep going through life pointing fingers, if its everyone else's fault, the person who ends up closed off and negative would be ourselves.

How we speak to one another (tone) has an impact of how energy is received, but sarcasm makes everything worse. Society now thinks it's funny to have a snarky meme, a comeback, an "Ooooh, he told you" moment; but it breaks down what little verbal energetic communication is going on. Children especially do not have a basis for sarcasm. I see many videos online of parents tricking their kids for a laugh and views, but what it does is breaks down trust and the barrier between what is real and what is fake. The ability to be able to distinguish between what is truth and fiction becomes difficult without a solid foundation of what is true, possible, physical, and knowing who you can and cannot trust. These actions, especially in childhood, severely impact the ability to discern right from wrong and severs the ability to access intuition.

How do you turn negative situations into wisdom? What kind of lessons are based on the choices you make or continually make in your life? The good, the bad, the difficult. Did you know you were choosing the wrong career?

Did you know you were choosing the wrong places to spend your time? Do you know who you should or should not be with? We try to make the best possible choices with the information we have at the time.

People do this in relationships at least once in their lifetimes. For some people, they continually make the wrong choice because they haven't taken the time to reflect and truly alchemize what happened in the prior relationship(s). For example, you continually choose someone who is beautiful, spontaneous, loves to travel, and experience new things in life. However, none of them have stayed. Why? It could be that your energy doesn't resonate with them. You may be waiting for the other person to take the lead, suggest a new restaurant, destination, or experience. That person may not be wired to stay in a long-term relationship. The qualities that you are attracted to change often. It normally doesn't happen that two spontaneous people can stay together because it would require them both to change at the same time. I'm not saying it's impossible, just not probable.

One of the most difficult things to accept is that we are choosing to be where we are right now. Say out loud, "I am *where* I choose to be" or "I am *who* I choose to be" or "I am *what* I choose to be," It can suck, because maybe we aren't where we would like to be or who we want to be, but we must acknowledge ourselves as we are right now. We are at the weight we choose. We are with the partner we choose. We are living the life we choose to live. This is how you begin to draw your energy into yourself. By taking the power of your choice, your free will back into yourself. You are drawing Etheric energy into yourself. Now that you acknowledge where you are, make a choice, any choice, to move closer to what you truly want.

Fear will keep you in a place you don't want to be. Fear paralyzes you by either keeping you in a relationship you don't want or in a physical place you don't want to be. Negative energy will creep into your mind by saying you don't deserve a better relationship or though virtue, "I can help save this person." Both are negative and thus deplete you. The same goes for a job. If the energy is not replenishing you, make a choice, any choice, to change. You will find out if your choice is positive or negative by receiving signals. Synchronicities of numbers or affirmations or what you are focusing on will start to be drawn to you.

You can have analysis paralysis, the fear of making the wrong choice, but do it anyway. Starting the momentum of exercising your power of choice draws power to you. If the first choice doesn't work, make another. Try to make the best decision you can, drawing energy into yourself and replenishing your energy. Your choices are up to you. Try not to ask for external input.

How you feel about your choice is important. We've been told not to listen to our feelings since we were young. Have you ever heard the baby-elephant story? Circuses will chain a baby elephant to the ground with a heavy iron tether staked to the ground. The baby elephant pulls and pulls, but because it is small, it cannot pull itself free. As the elephant grows, it stops pulling and stays where it is placed. Even though the elephant has grown and is strong enough to pull itself free, it doesn't, because in its mind, the chain is too heavy.

As we grow and learn, we become stronger. We can break free of the chains if we try. Enter in the concept of the power of limiting beliefs. Money (energy) isn't only for the wealthy. Energy is exchanged every day. As our battery capacity expands as we learn new skills, we have the ability to pull ourselves free and move on to the next level. We can take on a new task that will have greater rewards. Starting a new business, writing, painting, consulting, doing for ourselves instead of lending out talents to a corporation. We've all witnessed more internet millionaires over the last couple of years than I've ever thought possible. Why, because a direct connection has been made from the skill set of the individual to the end purchaser. We no longer need a middleman. We do not need a conduit using our energy for a company's purpose instead of our own. Through the Ether, the metaverse, the online community has been growing since the nineties. We as humans have developed a way to replicate ourselves and offer products and services without the oversight of the overlords.

Concept 12
Energy Transfer of Money

Money from your work, valuing yourself; valuing your time, efforts, contribution; holding your frequency, receiving money, receiving gifts.

What is frequency, and what does it mean to resonate with someone or an idea? We all know when we're enjoying a conversation with another person, when we have commonalities with work, life, or personalities. Finding the people we resonate with is part of this game of life. The speed-dating version of connections for business is called networking; however, most of us network wrong. Are you part of a group just for name's sake, such as a chamber of commerce, local charitable organization, or some other group where you go just to be seen? Like it's a check mark on your to-do list? Or are you making genuine connections? How much time do you give it? Well, how much time do you give to a person you're dating that you're not interested in and you see no potential? Damn. So many of us waste so much time going through the motions of connection but we're not actually connecting.

It is a good thing to help those that resonate and respect your frequency. You should not help those that drag you down, make you feel less than, or question your worth, especially when you are gifting your services to others. Many people have this issue in business with friends and family. If your friends and family are wanting the "friends and family discount" you need to implement a "no friends and family policy." If anyone is asking you to discount your services, they do not value your time, skills, or efforts and are best avoided. Remember where you receive your money from matters. Are you receiving money (getting paid) from a caring or, at least, neutral source? Does this person or business have control over you if you receive money from them? Are you doing things that make you feel uncomfortable in order to receive money?

In work, what you are offering must be able to be received by another. Is what you are offering solving a problem? Providing comfort? Teaching? If you are presenting without value, then you're just noise and your energy will not be welcomed by another person.

Giving your power away in exchange for money—I didn't realize this until recently. I was in an industry that was highly restrictive on what I could or could not say online. My speech was restricted to the point where, at the initial creation of online social accounts, MySpace (I'm old), Facebook, and other social media accounts were restricted. Heaven forbid I publish my own content. It wasn't until just recently that the controlling entities allowed people to have online accounts. I gave my power away because I chose an industry that I thought I wanted to be in. I realized that's why many popular "financial gurus" do not hold any professional licensing. To have their freedom of speech allows them to help more people and in turn make more money, for everyone in that exchange. If I had really stopped and evaluated what I was giving up, my freedom of speech, freedom of expression, freedom to communicate with others, I would have realized it wasn't worth it; and I probably would have written a book much sooner.

Do you work at a company that values what you value and believes what you believe? Or if you own a company, do you work with people that share or respect your values and beliefs? Some people say, in order to make sales you have to accept other's beliefs. Which, in a way, can be true. But what I'm talking about are core beliefs: family, spirituality, commonality in lifestyle and decision-making processes. And you know what? It's okay to not make a sale. Remember when I said that the Universe begins to present options to you when you take action? If you keep saying yes to those sales from people or companies that you really don't want, well, the Universe responds to your action, not your thoughts. So you will keep going down the path of what you really don't want, because you're saying yes to it. Are you being your authentic self? Do you have to put on a corporate plastic smile every day? If you are earning money with your false/fake self, then the energy/money you receive is also fake and will not fulfill you.

Choosing a job that matches you means matching what your core beliefs and values are with those in the company (co-workers) and the company as an entity (corporate values). What interests you? If you could sit and watch any show or read any book, what would you spend your time learning about? What holds your interest?

Once you have a subject you're interested in, find your tribe! Finding people who have common ground with you is the reason why social media is so successful. You become part of a group that enjoys watching, learning, or laughing about the same things. Having this tribe surround you in life will create a support system and make you feel like you're not alone. You will have people to share ideas with and support your thoughts on the common subject.

Do you feel worthy of receiving? This is often a loaded question that can bring up psychological trauma or generally bad feelings about ourselves. "Are we worthy?" is a phrase that carries a lot of weight. In our minds, we immediately compare ourselves with others we deem as our peers or the standouts in society. But it is a deeply personal question. Can you receive a compliment, a small gift, like someone buying you a coffee? What about a raise? What about charging for your skills in a manor that represents your value? Starting with a position of gratitude will help. This is more than just a Pinterest quote slapped on some shiplap. Waking up and being grateful for being alive starts the cycle. Throughout the day, be thankful for little things. That flower on the side of the road is so beautiful, I'm glad I saw it. The sunrise/sunset is so colorful—appreciate its beauty. I got the first parking space at the shopping center—awesome! Small things like this will reinforce the action that is happening, and Universe recognizes your appreciation and will bring you more of what is beautiful, lucky, and helpful into your life.

What is your feeling toward money? If you feel that money is a necessary evil, you won't have it. If you hold on to your money and do not share what it can buy—food, clothing, maybe a vacation with your family—if you are stingy, the Universe cannot flow into a closed-off space.

How Do You Value Your Skills?

If you are creating a product, search online to see if someone else is producing what you want produce. How is your idea different? How is it the same? How many hours did it take you to produce? For example, a person decides to hand make candles and can maybe sell them for $50. How many hours did it take to make? Two hundred hours, not including setting time but includes cleaning and packing to ship time, cost of materials and sales? Ten dollars (if you make 1,000 at a time). So 1,000 candles at $50 is $50,000. Materials (10,000) is $40,000, for 200 hours, or $200 an hour. You can afford to hire some help. That's a pretty good candle business.

Say you're making bars of soap, which is popular now. One thousand bars in self-production is quite a lot for anyone in home business or side passion. I've seen these sell for around $8, cost of materials being maybe .50 each. So 1000 bars at $8 each is $8,000, minus $500 for materials is $7,500. Can you sustain the lifestyle you want with this level of production and revenue? Even though the profit margin is high, the amount of volume you would have to do without some sort of mechanization and a heck of a marketing and distribution plan, may not be worth the effort.

When choosing a product, is it so individualized, specialized, or of a niche that it would narrow your sales or would being specific help you choose your ideal market? Are you able to replicate yourself/automate your process so you don't have to repeat the same thing over and over. Is hiring help an option? Figuring out *how* you want to make money is key.

Services

Replicate-able* can you repeat your process? (* since spell-check says *replicate-able* isn't a word). Products that require more production, more materials, labor, or machines to generate those "thneeds" that all people need (Dr. Seuss teaching supply chain) can make for a complex business model.

This can also happen in what is seemingly a service industry that delivers a product. The most recognizable is a franchise, for example an esthetics office, salon, medical practice, restaurant, etc. You will still need to factor in your time, effort, energy, and missed opportunities while working on your business such as missed time with your children or other activities.

This meditation is to help you discover your purpose and create your life plan by guiding you to choose a business that best resonates with you to achieve financial and spiritual alignment for success and abundance.

Below is a business meditation practice where you can ask yourself a series of questions to begin to discern what you really want in your life and what business model can provide the structure for what you're seeking.

*****

The following are questions every entrepreneur should ask themselves before going into business.

You will achieve the best results by participating in active meditation. There are meditations that ask you to focus and empty yourself to allow information to come in. This is a masculine form of meditation. The feminine form of meditation is to be filled up. We do this by asking questions of ourselves. Think about each question. The first time through this process, just sit and question yourself. See what information starts to come to you and make a mental note. If you feel so inclined, take notes. You will begin to feel yourself answering the questions. Listen to the answers. They are coming from what is resonating deep within your subconscious and from what resonates best with your soul.

After reviewing the questions start walking or doing a mundane repetitive task such as mowing the lawn, housework, or even driving. Listen to yourself in the silent moments. Instead of being completely relaxed, keeping your physical body occupied with a physical task will help bring ideas from the Ether and manifest them into the physical world. You will do this by writing down your ideas, thoughts, and impressions to begin forming the mold of the business and life that you want.

Question 1: What type of lifestyle do I want?

Do I want a glamorous, fabulous lifestyle? Do I want time with my family and have enough to be stable and take time for myself? What am I really willing to do, and what do I want my life to look like? Am I a night owl and like to work while everyone else is asleep? Am I an early riser and like to get my day started and mandatory tasks complete in the beginning of the day? What types of businesses lend to that type of lifestyle that I want? Do I have a bubbly personality, or am I more reserved? What location do I want to work from? Do I want to be out and about meeting new people? Or do I prefer meeting in a more professional setting such as an office space? Am I able to be productive working from home? How many hours a week would I like to work? How much time do I need for myself, my family, or other personal obligations? By keeping this lifestyle in mind, I will make choices congruent with the life I want, and I will not burn out with a business that does not match my biorhythm of life. I need to choose a business that is in resonance with my life. What type of lifestyle do I want?

Question 2: What problem am I solving, and am I passionate about solving it?

Every service or product has to be of value to my potential customer, or they will not buy what I am offering. Am I interested in selling nontoxic products? Am I interested in teaching someone a skill set I have a great deal of knowledge in? My business needs to be something I am passionate about. If I do not care about my product, then there is a disconnect. Then any potential customer will see I am being false. What is a problem in my life that if I fixed it, my life would go much more smoothly? Do I have the ability to solve that problem? Can I offer that solution to others? What problem am I solving, and am I passionate about solving it?

Question 3: Will I enjoy doing this task every day for years or the rest of my life?

Do I have a hobby that I am passionate about? Can I put more focus into my current side hustle to make the money I need? Test the business out first before I totally commit. (Many people are a fan of "burning the boats," but that is not realistic. A caterpillar doesn't just decide one day to become a butterfly, and the next day the transformation has happened. The same for the sun rising and setting. There is a transition that occurs over time.) As long as I am making steps in the direction I want to go, I will begin to change naturally, and my ideas can blossom into a fruitful endeavor by thinking, taking action, and repeating my efforts. What task will I enjoy doing to receive income for the rest of my life?

Question 4: What does it take for me to start the business?

Can I take a few online courses or start online with products and have a turnkey system available immediately? How much does it cost to start the business? How long does it take to make money in the business? Can I partner with others to spread the workload while remaining an independent entrepreneur? What about affiliate marketing? Do I know a lot of people or have thousands of contacts online? What products or services participate in affiliate marketing, and do I have enough of a following to convert to sales? Do I want to take a physical product to market? Do I know how to manufacture this product? Do I have the equipment necessary, or do I have connections to create my product? Do I have a sales tracking software or a bookkeeper? Can I create a website with integrated payment processing? Does the business I want have a high barrier to entry? Is it expensive, or do I have a huge learning curve? Does the business I want require a building, employees, or machinery? What about buying into a franchise? How long will it take me to get up and running? What does it take for me to start the business I want? (as you go through the questions not all of them will pertain to you. Some questions will hit harder than others, those are the questions you need to dig deeper on.)

Question 5: What type of income do I want? Residual/Recurring income or per sale?

This is not only the dollar question but the million-dollar question. Does my business require me to make constant sales (restaurant, real estate)? Or is it recurring (people purchase products each month, like a grocery store, health products, or rental income)? What about residual income? Can I make the sale once but continue to provide advice or coverage month after month (consulting, insurance, subscriptions). Is my business structured to create recurring/residual income? How is the business I want structured to make money? Is the way I am earning income in alignment with how I want my money to flow to me?

Question 6: Who is my customer?

Who do I feel the most comfortable working with? Who will benefit the most from my product or service? Are my customers a reflection of myself or are they totally different from me? Your customers should be similar to you, if not, your fake self could be showing up and this will create issues down the road. Who is my tribe? Who is willing, able, and ready to buy from me? Like goes to like. Am I familiar with who I want to sell to? Do I know their buying habits? Can I reach out to a group of potential customers right now? Who is my customer?

Question 7: How do I reach people who are willing and ready to purchase from my business?

Where will I engage my customer? Is my product able to be sold online? Do I need a store front? Do I have the skills and understand the technology necessary to be efficient? How will I find my customers? Do I know how to reach people? How will I interact with them? How will I deliver my product or service to my customer? Am I comfortable talking directly to people? Do I need to have someone else be the front person? Do contract work or partner with someone who does what I do not want to? How do I reach people who are willing and ready to purchase from my business?

Question 8: Do I want to own a job, or do I want to own a business?

What is the fulfillment of my offering? How much am I willing to work as one person? Do I need an assistant, a partner, or employees? Can I put systems in place to help me run my business efficiently? Can my business run for a duration of time without me? Can my business run without me while I'm on vacation? What if I get sick? What if I get sick for a long period of time or other personal obligations come up that pull me away from my business? Can my business run without my daily involvement? Many professionals fall into this job trap. Attorneys, doctors, solo entrepreneurs. Placing myself in a business that compromises the lifestyle I want by having to continually produce is draining. What business or business structure will provide me with a continual source of revenue? Do I own a job, or do I own a business?

The Questions:

What type of lifestyle do I want?

What problem am I solving and am I passionate about solving it?

Will I enjoy doing this task every day for years or the rest of my life?

What does it take for me to start the business?

What type of income do I want? Residual/recurring income or per sale, transactional?

Who is my customer?

How do I reach people who are willing and ready to purchase from my business?

Do I want to own a job or do I want to own a business?

Take time during the day to think about one or several of these questions. Be as completely honest with yourself as possible. If a business seems lucrative and pays you well, but you are working eighty hours a week, this will take its toll on your health and well-being. Choosing a business you are passionate about that supports you financially but also emotionally and spiritually will be more fulfilling and in alignment with bringing joy and abundance to your life.

Concept 13

What Is Stopping You from Moving Forward?

What one thing can you do today to move toward the goal you have, toward a more ideal life for yourself? It doesn't have to be big. It could be discovering a healthy meal that doesn't completely taste like cardboard and grass clippings. It could be reading a chapter of a new book or spending more than just a few more minutes with your kids at night before bed. It could be recording your first video. Writing your first chapter, even applying for a job you think is interesting or where you think you'll get experience are all small ways to move yourself in a positive direction.

When we are working past the job, past the career, and get to our passion, this is where the Etheric energy comes in. You begin to find ways to multiply your energy. How?

Say you charge $50 for your skills by sharing your knowledge. You decide to create a class, maybe a painting class, business class, whatever your skill is. Now you have people coming to you and instead of one-on-one, you present to a class/group of students. More than that, you create an online course that has now interacted directly with multiple people (add zeros as you feel necessary).

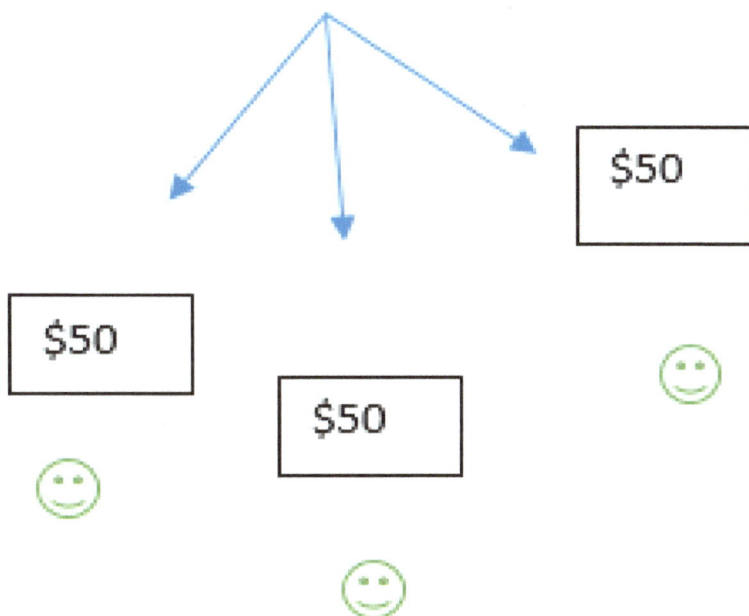

With creating energy once and generating a ripple effect, you have essentially multiplied yourself. "Go forth and multiply." Where have I heard that before? How about "Go forth and create abundance." This is what happens with a dream job, living the dream, finding your passion, sharing your energy. You have now manifested abundance.

What stops us from reaching this level? Leaking our Etheric energy prevents us from going from the daily grind to abundance. We leak our physical energy by being inefficient when we run errands or when we forget where we put our keys or our coffee in the morning. We drive to three different stores all in a day all over town; but if we stopped to plan for a few minutes, we would have saved time, gas, money and our energy. We all do it. Etheric energy is sometimes difficult to identify, or the leaking of energy may be a habit we don't want to break.

How do we leak energy? Social media, movies that have a low vibration, friends, family, coworkers who are time suckers, taking care of someone who isn't your responsibility.

How do we identify and get rid of energy traps? Get out of what isn't serving you. Wait, isn't positivity and attraction about serving others? Yes, but if your energy is drained, that's not good either. If you know in your gut you need to change jobs, not talk to that "friend" anymore, quit giving your alcoholic neighbor grocery money because he spent his money elsewhere. In the Gospel of Thomas, Jesus says, "Don't lie and don't do what you hate." Yet here we are. (The Gospel of Thomas was omitted from in the mainstream Bible.)

People will drain our energy, drag us down instead of lifting us up. Leaking energy adds to an unbalanced exchange of energy and focus. Congruency gets us to where we want to go. Continue to take steps in the direction you want to go. If you want to teach a course in making herbal remedies, but you haven't written the course, or maybe you have a few items down but keep stopping, you'll never get there. If you want to lose weight and in the morning you eat well but you end up having lunch with a friend and eating a bad meal, then "recover" by having a salad. You're taking two steps forward and one step back. Sometimes that one step back is a leap and not a step. A health coach I know said one burger doesn't make you fat, but one salad doesn't make you skinny either. It's the sum of our choices.

Unbalanced Focus—Unbalanced Energy Exchange—We are the sum of our choices

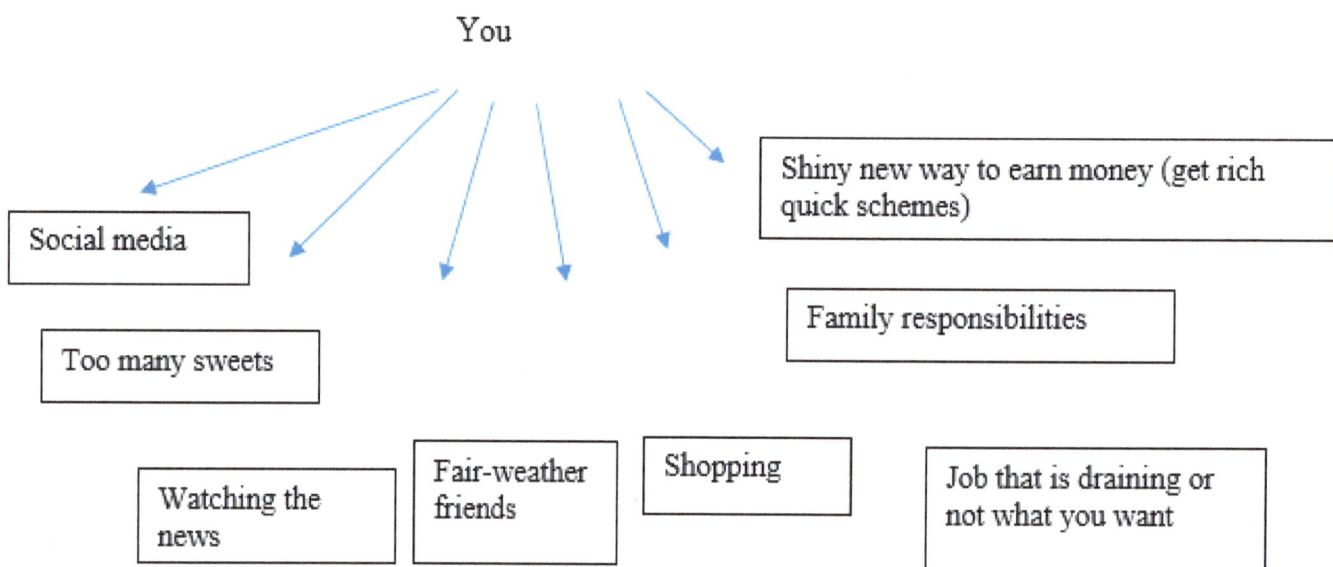

You

Social media

Too many sweets

Watching the news

Fair-weather friends

Shopping

Shiny new way to earn money (get rich quick schemes)

Family responsibilities

Job that is draining or not what you want

In addition to leaking energy many of us are also guilty of giving our power away. Much like my example of choosing an industry that created boundaries on my communication. Energy loss when we're interacting with others is usually the power we've given away. Do you really think your boss has power over you? If you think that, then it is so. Do you really believe you are worthy of accepting that promotion, then it is so. The power of belief will help you heal in a medical situation. It has been well documented that the power of belief can create miracles. With focus, belief, and intent, miracles continue to happen through the unseen energy of the Ether.

The higher your frequency, the greater the pull. This is what some call the Law of Attraction. The reason you are able to attract is frequency and magnetism. Your heart is an energy source with a unique resonance. When your frequency increases, the greater your "bubble" becomes.

Your frequency goes out into the ether and begins to pull toward you those that resonate with you, that have similar ideas, business partners, love partners, your "tribe." It can also help manifest where you want to live, how you want to design your life.

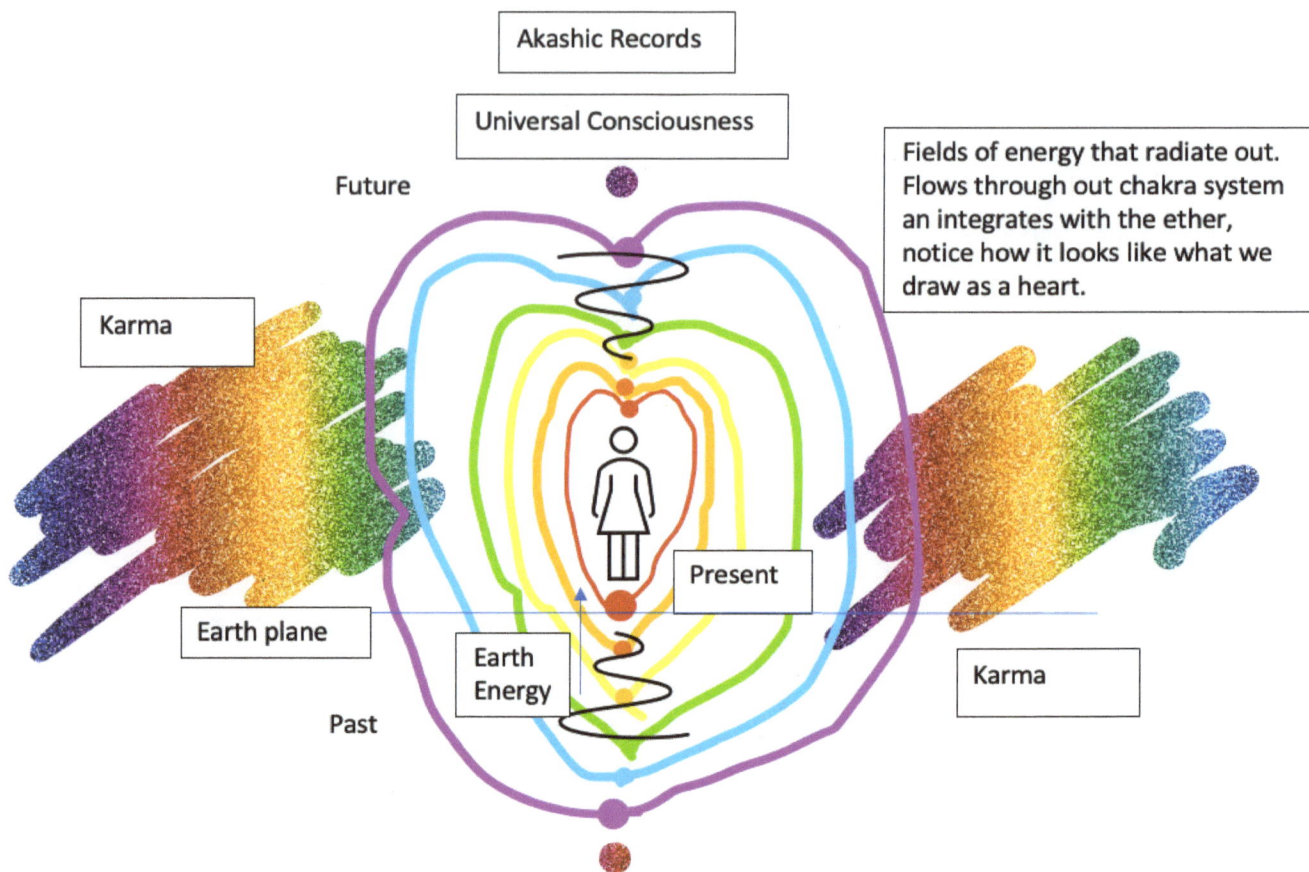

Akashic Records

Universal Consciousness

Future

Fields of energy that radiate out. Flows through out chakra system an integrates with the ether, notice how it looks like what we draw as a heart.

Karma

Present

Earth plane

Earth Energy

Karma

Past

Energy flows both ways and through the center of the body. You will know what positivity feels like through the fruit it bears, what feels easy and effortless as you begin to reap rewards. For example, if you enjoy making food, you open a restaurant, plan, prepare, decorate, and market (so you do the work); and customers come flooding in. Your customers feel your energy. They taste it in the food and tell their friends, so the energy spreads through social media/word of mouth. This energy transmission happens through the Ether.

What is high vibration? Frequency of the energy of the earth, plants, the sun; morality, compassion, love. What puts you in a higher vibration? Perhaps music of a frequency that resonates with you. What is low vibration? Taking on someone else's burden. You have to get to a place where you are not afraid. If you are afraid to make any change in life, or you will never get out of the cage you've placed yourself in.

The Book of Thomas verse 20 says, "If the master of the house knows a thief is coming, he will be vigilant and not allow the thief to break into the house of his kingdom and carry off his goods. Thus, you should be vigilant toward the world. Strengthen yourselves with great energy or the robbers will find a way to get to you."

We are all reflections of the interaction of the Universe. Universal Law is seen in all things, pulling from what our eyes cannot physically see but what we can feel and interact with.

Concept 14
Why Do We Keep Making Bad Choices?

Reasons why we keep making bad choices:

1. What we're choosing cannot be because we're not on the same frequency as what we want. As shown in previous concepts, your thoughts, your words, your actions must all be in alignment with what you want. If you want to attract a partner, you must be your authentic self and also remain in a positive thought. Show loving and caring action toward that person. I'm not saying be a stalker or buy them needless gifts, but be present in conversations, listen with the intent of getting to know that person better. Your vibe of caring will come across. I can't say it will be matched by them, but this will give you the best chance of having your feeling heard.

2. The person we think we want is not truthful and can manipulate the frequencies around themselves and you to get what they want (narcissists). Much like a spider catches their prey by surprise because the prey cannot see the web, we as humans become trapped in others' webs. This is where we get the term "web of deception" or "web of lies." We must become better at detecting this "invisible" web. The higher our vibrations are, the better we are able to see the web.

3. We haven't taken the time to reflect and understand what went wrong. Why didn't I get the job, relationship, house, etc. How was I acting? What were my reactions to the situation? We can only control ourselves and our reactions to the happenings in our world. If we did all we could, we have to be open to the possibility that the Creator/Universe has something else better in mind for us.

4. We have programming that blocks our ability to interpret signals (intuition) correctly. If this is the case, we have to go back to the basics of emotions to understand what truly makes us happy, sad, fearful, or safe. Go "back to the beginning, Vizzini" (*The Princes Bride*) and recognize the root of our feelings.

5. We have not encountered this experience before and therefore have no basis for how to interpret the situation. This goes back to intuition and going with our gut. What does our gut say about the situation? (The gut is the second brain, you know.)

6. We expect others to change. By expecting others to change, that puts all of the power outside of yourself and faith in another person. It is one thing to put faith into your Creator/Universe, but placing expectations on another human being will only break them and cause you to be disappointed when the outcome is not what you wanted.

7. Hoping/praying for events to happen on their own without any action or intervention. People confuse the terms "hope" and "faith,' and use them interchangeably, but they are not the same.

8. All of our thoughts and actions are incongruent. All of our choices, thoughts, and steps must be in the same direction.

In life, a lot of people need to learn from the school of hard knocks. We have to experience the heat of the flame instead of being told the fire is hot. Once we can gain perspective through experience, meditation, active thoughts, or having conversations with those we trust, we can make choices that are in our highest and best good.

Some of us are damaged purposefully from the matrix/regulated society. There was a program running in plain sight, designed by those we gave our power away to, who wanted to control and manipulate our energy. Everything from TV programming to the way school (especially elementary school) was conducted to looking at social media and having them tell you what you're supposed to think and how you're supposed to feel and act on a certain topic. Many of us experienced suppression by parents or those in "authority." Instead of being caretakers, they took on a role of disciplinarian and conformed you into one way of thinking and being.

In order to "undo" this spell or programming, you have to get back in touch with the basics of your own feelings and emotions. There I something called the emotion wheel. It is used by psychiatrists with children so they can visualize what emotion they are feeling.

We're taught to repress our feelings of not wanting to be close to someone out of social graces. Well, Uncle Ned smells funny and Aunt Sue hugs you just a little too long. (By the way, most sexual abusers come from a family member or close family friend.) We're taught to clean our plates. We're not taught how to read food labels. Both are damaging to our diets. In school, we're taught to sit still for hours on end, repressing our urge to go outside and play, dance, sing, and make noise like children are supposed to do. Instead, we get scolded, told to sit in a corner by ourselves, isolated from the others if we "act out." We ask our parents, "Why?" "Why is something done a certain way?" and we get back, "Because I told you. Don't talk back to authority." It's not about talking back, it's about two human beings interacting who happen to be born fractional moments of a lifetime apart on the universal scale of time.

Energy of Perception

Humans have a natural preconception of others. No two people are alike. Two men can dress in suits, both Caucasian, styled hair, clean shaven, fancy cars. Do they both carry the same attitude toward others? If you continually perceive others with your lens of experiences, you will push people/opportunities away from you with your misplaced energy. What? I thought I was supposed to go by my experience and my gut. While experience is valuable, having the ability to judge each situation independently of others is what will bring you better decision-making capacity. Having a bubble of protection around you is one thing, but pushing others away that otherwise have a frequency the same as yours can create missed opportunities. We do not know the form or shape our blessings will come in. It can be in the form of a person you would never normally interact with. It can be in a location where you normally don't travel. Our lives are the sum of experiences, interactions with others, and our choices. If we limit our world to a certain color, flavor, place, etc., we miss out on new experiences, growth, and opportunities.

Energy of Revenge

The best revenge is to alchemize the pain from an experience, learn from that experience, and move on with our lives. Understand what happened and acknowledge it. Some experience pain from the depths of hell through physical and emotional abuse. Others experience love being used as a weapon. This kind of pain needs to be acknowledged. Using the wheel of emotions, figure out exactly how you are feeling and then what you can do to shift yourself out of that negative space. You may never be 100 percent whole, but you need to gather as many fractals of your energy/heart/spirit/soul/mind back as possible.

How to know when you've made the right or wrong choice? By slowing down your decision-making time and assessing what core feelings you are experiencing with each choice you make will help you process your emotions. What are your options? Does one make you feel happier than the other? Does one make you scared? Are you able to visualize yourself in the next ninety days with either choice? Even simple choices can begin to add up. Something as simple as, what's for lunch. A burger and fries or a burger and fruit? What about a salad? Who really even likes salad anyway? Maybe you just haven't found a salad that you do like or healthy choices that you do like? Especially in the US, we're programmed: burgers, fries, some form of chicken shaped into a nugget, pizza, pasta, or a sub. That's about it for food choices unless you go into a restaurant to sit down on your lunch break. Mix it up. Discover something new in the way of eating healthy. Does the chocolate make you happy? Ninety days from now, if you ate chocolate every day, would that really be the best choice?

The more you dwell on a bad decision, the more energy will leak away from your body. Just make the next choice, get yourself out of the bad choice hole, and move on. When you begin making choices, assess after each choice how you're feeling and the outcome of that decision. You may have continually been pouring money into a business that wasn't bearing fruit. The same fruit-bearing goes for relationships. Does your partner make you feel valued? Do you feel more rested, happier, energetic, looking forward to activities? By asking questions of yourself, you become more aware of the truth of what is going on in your life. I'm not saying to interrogate yourself after every interaction. Just a question or two throughout your can help bring clarity.

Emotion Wheel I

Uncomfortable Emotions

- Sad
  - Lonely
  - Hurt
  - Disappointed
- Scared
  - Anxious
  - Powerless
  - Overwhelmed
- Angry
  - Bored
  - Jealous
  - Annoyed
- Embarrassed
  - Ashamed
  - Excluded
  - Guilty

Comfortable Emotions

- Happy
  - Caring
  - Grateful
  - Excited
- Loved
  - Respected
  - Valued
  - Accepted
- Confident
  - Brave
  - Hopeful
  - Powerful
- Playful
  - Creative
  - Curious
  - Affectionate

HumanSystems
Healthy Systems | Human Wellbeing

www.humansystems.co

What Seeds Are You Sowing?

Here is a small bit of large insight and may actually be a surprise. Some, if not most, of the seeds you sow are for someone else to reap the fruit. Excuse me, what?! The seeds I'm sowing are for someone else's harvest? Well then, what the hell am I doing? People say what goes around comes around, and this is why. This isn't energy leaking; this is a form of energy exchange. If you sow discord, strife, hate, and drain other's use of energy, that is what you will receive. If you are asking for handouts and expect people to do things for you then you are syphoning others' energy from them. If you sow kindness, love, giving, then you will come across someone else's seeds that are now bearing fruit. This is the ultimate energy exchange.

Weeds will grow in fertile soil. Nothing will grow if the soil is out of balance or depleted. If you do not pull what you don't want to grow, your garden will be overrun with weeds, and your efforts to bear fruit (or vegetables) will be choked out. If you pull only the surface of the weed, the roots will regenerate the problem. The longer you leave the weed, the deeper the roots will grow, and the more difficult it will be to remove.

The barren path, if your frequency is low, will bring others down around you.

What do you want to grow in the garden of your life? What are you planting? What are you pulling out? You ever notice how you don't have to plant weeds? They just find their way in. If your garden/life is not bearing fruit, you need to figure it out. You alone. People can help you water it, fertilize it, and share in the responsibility; but you cannot give up your responsibility and leave the growing to chance.

For example, a mother pours into her son. She gives of her energy in exchange for love from the child. Thus, the child grows up knowing love, knowing how to treat others, knowing what it is to receive kindness. A father shows how to offer gifts and skills to the world through example of leadership, kindness, interaction with others. This sowing into the seed, which is the child, becomes the man, who is the fruit for the woman he will marry. (All energy exchange between the parents' gifts for the child's love is interchangeable. A child can see both parents' work ethic, affection for each other, their own personal value to the family, and how to appropriately share gifts and skills with others.)

You multiply the seeds you sow through the interaction with those of the same vibration. This is why it is important where you live, how you live, and who you interact with.

The vibration starts at your core. At your core, you are dense. At lower density, think of your emotions being packed into a small area. This is why people who have a lot on their mind feel "weighed down" or feeling down. As your vibration increases, you feel lighter, less dense. I feel "free" or "like I can fly" or "on cloud nine." In the drawing below, that would be a very high-level of vibration.

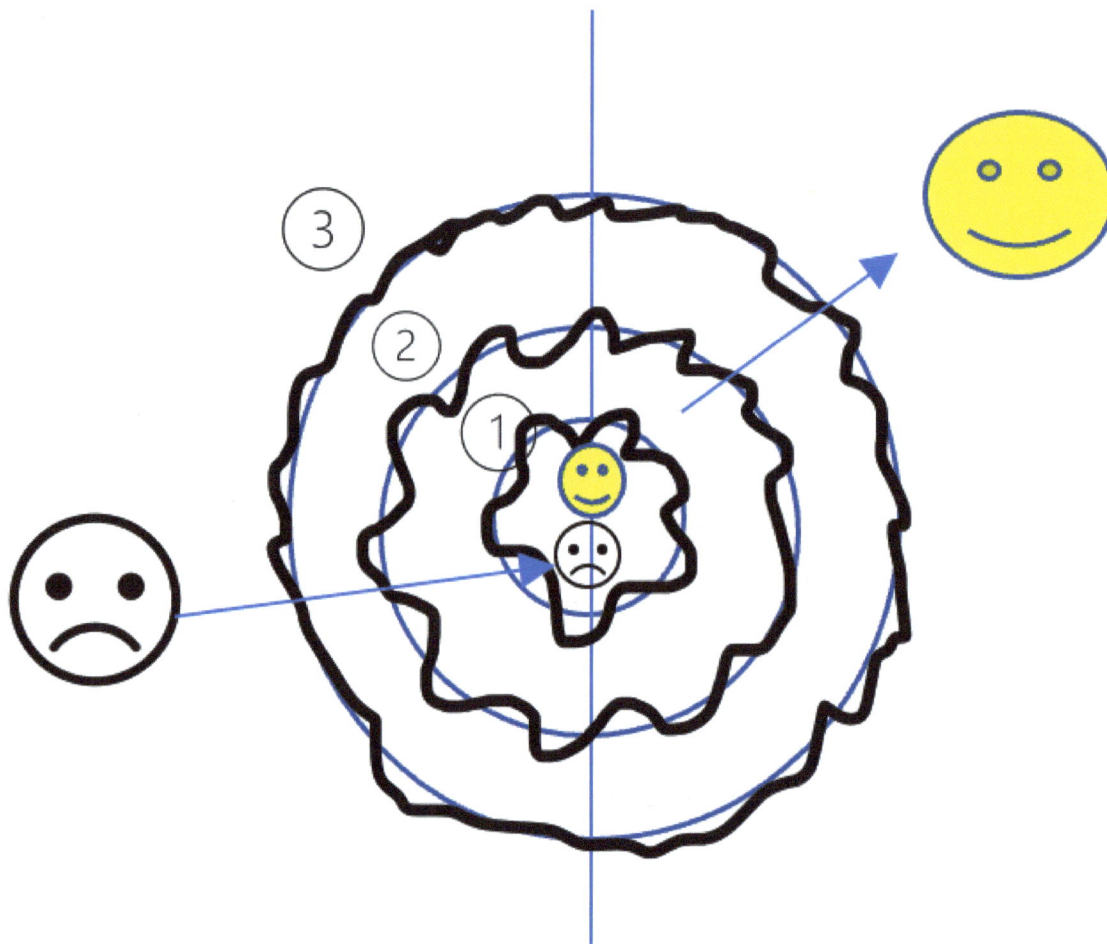

This is how positive people are overtaken by the negative. This is also how empaths work, by absorbing other people's negative energy. If someone has a low density/low vibration and are dealing with a large amount of negativity, it will bring others down, even if the other person is trying to create a positive bubble. You may have friends, family, coworkers like this, where you don't dislike the person but their aura is so toxic that you literally can't afford, energy wise, to be around them. They will begin to syphon off your energy. Whether either of you are conscious of it or not, your energy fields will try and tilt to make the interaction at least neutral. When you hang around someone with a positive vibration, they lift you up. This is how church sometimes works and sometimes doesn't. If people are genuinely loving and positive at the church you go to, your energy will increase. However, you cannot fake positivity. If people at your church are actually negative, they will bring you down. The plastic smile and fake nuances of churchgoers or other "positive gatherings" (think family at the holidays) can really do a number on our energy.

The key for those who are energy generators or for those who want to manifest is to push through the negativity around you and to be in high enough vibration for what you want to pick up, a signal that resonates with you. Are you beginning to get a glimpse into the Ether and to how much higher your vibration has to be in order to cut through the crap?

Here is another illustration of your vessel, your container, your body:

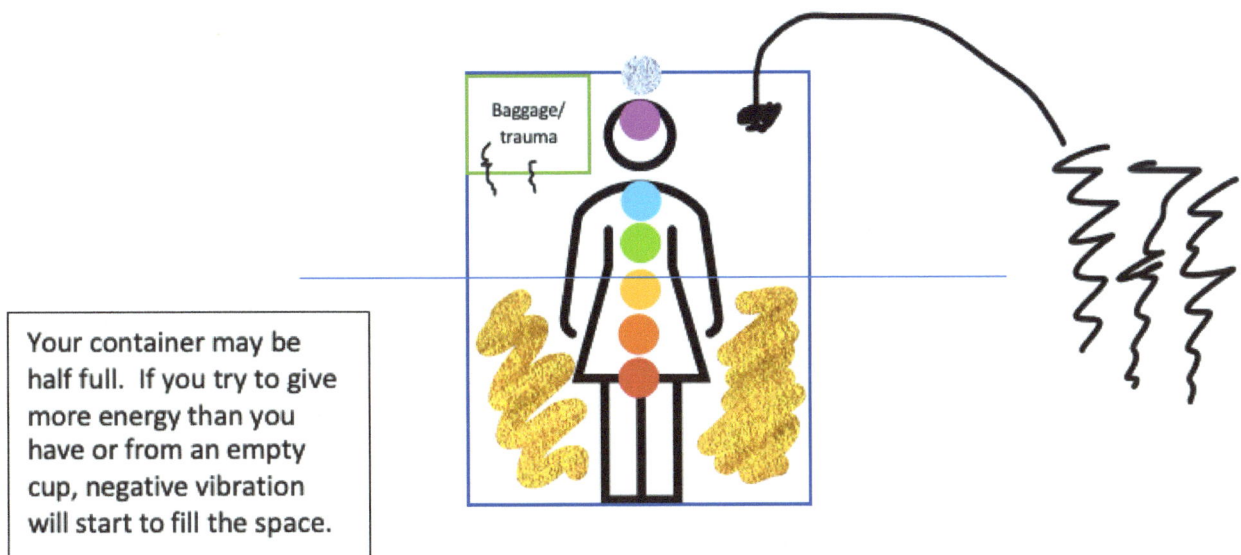

Baggage/
trauma

Your container may be half full. If you try to give more energy than you have or from an empty cup, negative vibration will start to fill the space.

Service to others does not mean depleting yourself. You cannot give from an empty cup and barely from a cup that is half full. If you are not abundant in your energy, the energy you are giving in a depleted state carries with it the opposite effect. You will begin to syphon. It is like wrapping an empty box. When the person you give the gift to opens it, their energy goes into the box to fill it up, thus balancing out the energy exchange. This is an odd sort of equation, like giving a negative number.

I'm giving you $25 for your birthday; however, it carries with it negative $5 for my lack of energy. What your intention to give (from the Ether) sets the balance of units for the equation. That person on the receiving end will unknowingly give the other $5 out of their emotion, thus leaking energy. This happens with time spent with people who you either feel obligated to be around, (could be family, could be that annoying coworker) and you feel drained when you leave. What about as a parent? We get tired of our day to day routine. Exhaustion sets in and nerves are frayed. What kind of attention are we giving to our children in that state? What about your partner? The Etheric balance gets its "pound of flesh" either way. It's like a quantum machine that only balances the outputs. If, then…if, then—repeating in each Etheric energy exchange for all eternity.

Some examples of actions that deplete us are: going to a job we don't like, being around negative people, or people in a sad and depressed state. Those who are complaining about everything or being in a total-victim attitude all the time perpetuates an aura of negativity and therefore are manifesting of the very thing they are dreading. This can be a difficult balancing act for those who work in public service, such as child advocates. They find joy in helping children but are exposed to the extreme filth and perversion of the lowest rung of humanity. They save lives every day, but what does this do to their psyche? They do not want to quit their work, so how do they heal? What do we do when we need a break? The term self-care has come about over the last decade or so. Taking the time to do what we enjoy, such as taking a salt bath, relaxing in a hot tub, washing our problems down the drain in the shower, taking a nap. These are all ways we "lift our spirits." Funny how all these old sayings keep working their way into my head as I'm writing. It's like they were based in a time when people naturally understood Etheric Energy Exchange.

There are actions that give us energy and actions that take it away. Gardening, hiking, being out in nature can replenish out energy, just as being with loved ones and friends. Yes, for the moment it can be tiring, but overall, it adds to our container for joy and comfort. Actions that deplete us are going to a job we don't like, taking classes we don't like, being around people we don't like. See a pattern? It is also being around people who we love and care about but have a depressing or negative outlook on life. This could be a spouse or a parent. It is a difficult balancing act being empathetic ("empath" is in that word) and giving of your energy without them being a succubus and "draining the life out of you." This can also be a difficult balance being in public service, a first responder, social worker, or an attorney that represents children. People in public service do so because of the joy they receive from helping others in their community, but what does continued exposure to intense levels of trauma do to a person's psyche? It can cause fragmentation within the person, creation of a different persona to handle the intense situations, and then when they get back home, removing that layer of protection and putting it away in a box in storage somewhere in your container. What can you do when you get home after being exposed to trauma or an intense situation? Turning to drugs or alcohol, smoking, or other negative stimulus are ways that negative energy seeps out of that box you put the trauma into. Those boxes you store the negative energy into are your organs. I will talk more about this later.

One of the most confusing and difficult spirals of negativity is the giver who has pushed themselves to the limit. We may have met this person, perhaps it's the "Karen" who in her own mind thinks she is doing so much for others. She doesn't realize that her attitude and passive-aggressive nature is spewing negative energy on everyone. The person receiving the gift may in fact receive what is necessary: food, clothing, some physical donation. The people around Karen, mostly the other volunteers, receive her abrasive, rude, short-tempered exhaustion as her energy body is desperately trying to fill itself in order to keep from spiraling out of control. This person's circle of friends seems to shrink, year after year. Karen is invited to do less and less. Her children do not want to have anything to do with her after they have the ability to rely on themselves. This is fake helping. If you can catch it early enough, if you're noticing some of the traits within yourself, this is a sign that perhaps you have a box of negative energy that you need to unpack and release back out into the Ether.

There are people who overthink things, causing analysis paralysis. They pray and do nothing and feel like they are in a state of suspended animation, not moving forward. This is also a universal law where nothing stays in state of neutrality. You are either moving forward, or you are slipping behind. Your equation is either positive or negative. No one stands at a zero state of neutrality. Even the monks in their meditative sate wish for the highest and best good, not neutrality. You can have packets full of seeds on your table. You can pray over them, energizing them with your intentions. But if you never plant them, water them, and tend to them, nothing will ever happen. If you don't at least start something, anything, you will have a 100 percent failure rate, and you will starve. Physically, emotionally, spiritually, if you do not plant your seeds, interact with the world, pour your energy in and receive energy in return, you will not have abundance in your life.

Hopes

Hopes are the worst. Hoping is knowing something bad could happen but you hope it will be all right. Hope is drenched in fear. Hope is when all else is failing and you are praying for a miracle. People interchange hope and faith, but they are not the same. Hope and preparation for an outcome are not the same thing. "I hope she likes me." "I hope I get the job." Hope is a mindset of giving your power away to chance. Hope is *not* the Law of Attraction. Hope is *not* any Universal Law. This mindset is *not* how you manifest what you want.

If you are not driven, certain, focused, and vibrating at the same frequency as what you want, you can hope all you want, but you won't get it. No matter how much you pray or put intentions out there. The Universe just doesn't work that way. The Universal Law is that actions must be met with actions.

Faith is knowing you will be all right even though something bad is happening. Faith without works is dead. Who even said that anyway? But they're right. Faith, pray (set intentions), and act. Belief is the dominating power of the Universe. If you truly believe in yourself, if you believe in your actions, and clear in your intentions, your energy will attract what you want with laser precision. Belief is what causes miracles to happen. Miracles are outcomes so far from statistical probability, that they are almost certain not to happen, but they do.

Will You Get Out of Your Own Way?

If you really look at what is holding you back—it's you. Your choices: "Should I eat this cake or have this salad?" "Should I create my marketing plan and implement it or think about it some more?" "Should I confess my feelings to the person I love?" The choices can be more serious such as upholding obligations. For instance, you may be caring for your aging parents while raising your children. The sandwich generation. Sounds yummy but it isn't. You're taking on more responsibility than you can handle, and your energy is what becomes depleted. We use our responsibilities as an excuse to not take care of ourselves or to let other tasks fall through the cracks. We may even don the crown of victimhood. Woe is me, I have to do this. I'm so pious for helping when I can barely help myself. Please, for all of our sakes, stop taking on more than you can handle. If you honestly ask for help from Source, you will get it. Not "God give me more energy so I can do all these things." The proper ask is "God/Source/Universe, I cannot handle all of these tasks. Please help me so this situation will come out in the highest and best good." The Ether works through you. You do not become the Ether. You by yourself are not the solution. Such as the electrical currents running through the circuits in our house. The circuits and wires are not the electricity. The electricity flows through them.

How we talk to ourselves during our choices also has a huge bearing on our situations and outcomes. Listen to your own words and thoughts. Pay attention to what you're saying to yourself and others. Would you talk to your loved ones the way you talk to yourself? If you really stopped to listen how you are speaking to your loved ones, is that really what you want to be saying, or are you grouchy and tired and need some self-care?

All of these scenarios stem from choice and also how we handle the outcome of the choices we made.

The getting out of our own way and congruency of choices manifests as a quite an Etheric dance.

Dieting is a little easier to see the interaction with Etheric energy (choice) and physical manifestation.

Be healthy

Day 1 +1 workout +1 smoothie for dinner +1 sleep = +3 in positive direction

Day 2 +1 workout -1 party with friends and had 5 drinks = 0 progress

Day 3 + 1 slept in − 1 stressed about eating badly +1 swam at the pool = 1 in the positive direction

Okay, it's not exact science, but you get where I'm going. If before we make our choices, consciously ask yourself, "Is this placing me in the direction I want to go?"

If want to start a new business. The key is to *start*. Take the idea from the Ether and do something, anything. At least one activity that will put you in the positive every day.

There is no division between personal and business life; we are the same being 100 percent of the time. It's time to start acting like it.

Now that energy is coming in, how are you designing your life? Your space? Are you able to receive? Create a list of earthly fulfillments and a list of spiritual (Etheric) fulfillments:

Happiness, health, loving space, food, travel to visit family and friends, travel to spiritual places, wearing clothing that makes you feel good, decorate your house like you want, etc.

People live their lives striving to meet goals. Setting expectations is different from having goals. Life is a series of choices. We are the sum of our choices. The product of our habits. (Product in math is multiplication. Our habits multiply our actions because we do these without thinking.) If any of this equation is out of alignment, meaning in the negative by the choices we've made, we've stood in our own way. The choices we make we must be accountable for.

There is no neutrality, but there is congruency. You are either gaining positive or losing negative. There is a balance to the equation, to each exchange of energy, but our containers are continually flowing in one of the two directions. Why? Because we are alive and do not live in a static environment. So there is 100 percent of energy to fill up your cup/vessel/container. It doesn't matter how positive or negative; the container will fill until its unit capacity is reached. Once the container is full, the surplus energy will then be thrown off. This is "hitting your limit" when negativity overflows *or* "my cup runneth over" when it's positive. "Bursting with joy," "Blow my lid"…the list goes on.

Here is another illustration of your vessel, your container, your body:

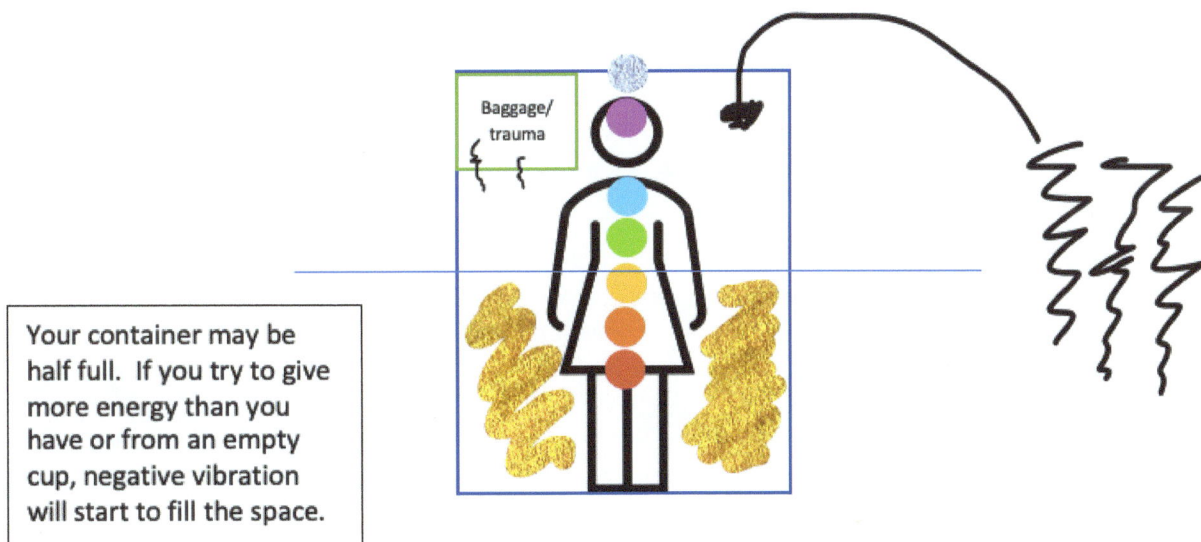

Baggage/
trauma

Your container may be half full. If you try to give more energy than you have or from an empty cup, negative vibration will start to fill the space.

We live our lives striving to meet goals, setting expectations, and when those expectations are not met, we are disappointed. Time after time we are let down. Life is a series of choices, and we are the sum of our choices.

Pay attention to the world around you. Create an awareness, not a crazy hyperawareness on the borderline of paranoia, but consciously aware. We can refer back to previous chapters where we recognize we have to live in the world as it is, while we manifest and work on bringing our future selves into the present. "I am a nurse working on changing health care." "I am a teacher working on accounting classes." "I am currently overweight. I am working on healthier choices." (Sometimes acknowledgment is painful.) Again, acknowledge where you are, where you want to go, and where are you *actually* going! Your words and actions must be congruent, moving you always in one direction. It helps to visualize your choices. We looked at this concept earlier on a linear equational format. Now let's see what our intention and attention looks like.

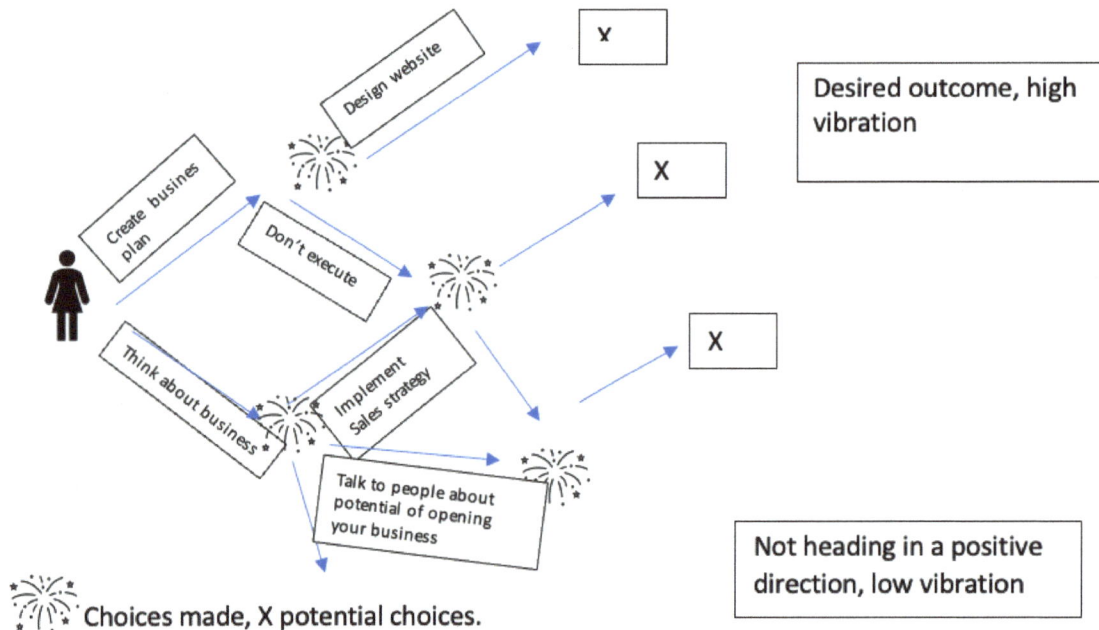

Choices made, X potential choices.

The difference is actionable items: designing, implementing, doing. You may have an awesome network of people and if you're a sociable, put that network to work generating energy for you. How network marketing became popular is having a product your tribe wants and then sharing your knowledge with others. If you get a queasy feeling in the pit of your stomach over MLMs or network marketing, why are you even attending networking functions in the first place? Why would it matter if anyone knew about your business? Why even build this web of relationships? If you're feeling stuck, draw out your most recent decisions and see where you're headed.

Energetically, physically, spiritually, mentally, you're okay most times; but is it a fight to stay focused? To stay on a singular path day after day, especially when no results are coming or you cannot see any progress. How do you know you are supposed to keep going? Is the path you have chosen the right one? Is your discernment off? We would like to give ourselves a time frame, a completion mark to measure our progress, to say this is working, keep going. But time is just one of many factors, some of which are outside of our control. If we are called to follow a certain path, make a product, create art, write a book, stay in a certain career, pursue a certain goal or a relationship, there is not only one person involved in your progress. Other people in the physical, interacting with the Ether, are involved in your outcome. Other people need to serve their purpose on their time in order to turn their wheel in the clock of life that in turn affects others. We understand this when we say, "It wasn't my time" or "The timing wasn't right" or "The stars didn't align" or "Time passed me by."

What we are really looking for are synchronicities. *What is a synchronicity?* There are quite a few books written on just this subject, and honestly, I haven't read any of them. This is my personal interpretation and experience of synchronicity factors. How to know what path to take means being aware of what and who is around you and noticing repeat occurrences. For example, an event, a location, or a person can be introduced to you by several different people at several different times. You could also notice a particular thought recurring in your mind or a message from a song on the radio, an advertisement, billboard, etc. The key is that you need to notice it. In order to pick up on synchronicities, you need to be present in your life. Guess what you're carrying around with you almost 100 percent of the time, a big 'ol distraction disconnecting you from the Ether. Put the phone down and pay attention to the world around you.

If we do not have a heightened state of awareness of being present in our conversations, by actively listening, observing what is around us, we are *missing* universal cues to guide our life. This is not to say you need to go all ninja with your senses or have perfect recall with your conversations, just start to look, listen, and observe the world that you are physically present in. The Ether will naturally seep in and start showing you what choices to make.

Now, how does this happen? How do we pick up on synchronicities? Well, by a couple of different methods. *First,* it could be a gut feeling, a pang in our heart, or a gasp of our breath. It can be either good or bad. It can happen if we find someone attractive, and we are looking to meet someone new. That someone could walk into your office one day, or you could pass them in the grocery store, and you just know that you have to talk to them. The feeling can go the other way by sensing danger or when a person or situation makes you feel uncomfortable. This is usually an immediate message from the Ether and causes you to make an immediate reaction.

The second way (and there three ways I'm listing here) is repetition from people or visual cues. You may or may not have an initial thought about the person, place, or thing. Maybe you've wanted to go on vacation but have been undecided about where to go. You first see an add on your phone (one, because you always have it in your hand, and, two, because the Google listens to everything you're thinking anyway), and you see the beach. Later that day, you're talking to a coworker, and she mentions she's just come back from the most fabulous beach vacation at XYZ beach. Awesome. Then as you're driving home, you hear on the radio an ad for XYZ beach, and its only $199 for the weekend. You started with a vague idea. The Universe got a little more specific on the second round, and then by the third round, it had it pretty much scheduled for you. Most people though may only start paying attention at the radio advertisement. If we take our time each morning or evening to take inventory of our thoughts (meditate), we may have been more aware of the synchronicities during the day.

Synchronicity of numbers/visual cues: many people see repeating numbers when a message is being delivered from the Ether. This can be repeating of the same numbers such as 11:11 on clocks or 4:44 or 12:34 or it can be a seemingly random set of numbers; but you see them all the time. Sometimes daily and sometimes even multiple times a day. So what the hell does it mean? In my personal opinion, this is what I've deciphered:

Ones: you are making the right choices and your life is on the right path

Twos: pay attention to a divine message

Threes: are synchronicities with other people

Fours: changes are happening and to make sure I'm being present.

Fives: you are understanding and in resonance with those around you or that there could be some conflicts or some decision you are 50/50 about that you need to make.

Sixes: are attraction and working with others, sometimes government or business relationships.

Sevens: be on alert for conflicts.

Eights: are spiritual, meaning you need to spend some more time in a meditative state, asking the Ether questions about what's coming next. Opportunities are opening up.

Nines: nearing completion.

Tens (1010): a completion of a cycle.

If you have some thoughts on what numbers mean for you, please let me know. Also, numbers can be personal, such as an address you lived at near a friend who passed away, a birth date, anniversary date, etc.

Synchronicity is not predicting the future. "As for the future your task is not to foresee it, but to enable it" (Antoine de Saint Exupery). We are honing our skills to enable the Ether to have a clear conduit to bring us what we need, from our thoughts into reality.

Third, you *physically* may need to move yourself. You could have perfectly caring parents and friends. You have a strong support system and love of your life; but if you are wanting to open a gym, start a farm, teach yoga, your current personal connections may not have what you need to learn these skills and *you must find your tribe.* Finding people with the same interests, passions, and goals as you and interacting *with* them, not seeing them as competition for your business, will help you grow. Another physical attribute to consider in order for synchronicity to happen is, say you do want to run a farm. If you live in a city, maybe you can urban farm, but you will need to be in a physical place to make that happen. If you have a dream of opening a restaurant on the beach, well, you need to be on the beach. Don't play baseball on a basketball court and don't play football in a football stadium. (Ball joke, I think you get the point.)

Having people and your environment be an intermediary of dropping Universal messages through the Ether to you is crucial to your success as a human being. Yes, it's that weighty. Have you ever heard your life will be like the five people you spend the most time with? Who do you spend the most time with (I'm about to lose my tact so please forgive me but…)? Your broke ass friends when you're trying to move up in life? Your family who questions your every move and doesn't believe in the business you're building? Your bar friends? Your wrong-side-of-the-tracks friends? People who have repeatedly made bad choices in their life? Do the people you surround yourself with have the life you want? If not, you *will not* be able to pull from the Ether what you need in order to pull yourself out of your hell hole. (Sorry, someone needed to hear that.) It's like trying to get electricity into your room, and the light switch is broken. The energy is out there. It wants to get to you. Your room is wired for it, but it can't get there because it's blocked by a faulty switch. Your conduit to the Ether is blocked because the energy is not flowing through who you are surrounding yourself with.

Concept 18
Intelligence and Intuition

Intelligence and skills both manifest themselves in the physical though your actions. People can see and experience your skill set. People can witness how smart you are by demonstrating problem-solving, innovation, and knowledge. Intelligence and skills are innate abilities that are nurtured by your environment, for example, by reading books, interacting with people, or being in nature. You can be born with a high aptitude to learn and process information, but you still must be in the right environment to fully manifest your abilities.

Intuition comes through meditation, asking yourself questions of your surroundings and asking others what their opinions are. Keep in mind that the outside information you receive, from your parents and teachers, are their interpretations of the world that should be brought into your mind and intuition for processing to demine if what you're being told is, in fact, the reality of the situation. People will see an event, a situation colored through different perspectives and interpretations. If you can get down to the actual actions that were done, the actual words that were said, only then can you make your own interpretation. This can be difficult, especially with past events getting only secondhand, thirdhand, or researched information. When this happens, a large amount of reading and researching on your own must be done to come to your conclusion.

How do we use our intuition with our intelligence and skills, to determine who we want to have in our lives? How do we know when a person resonate with us? When we meet someone for the first time, we assess them first on their physical appearance, but what else actually happens? Something we don't even see, but we can feel if we are able to tap into our intuition. We feel out their aura, their energy vibes that they are putting off into the Ether. As in the illustration in the previous chapters, are you vibing with them? Or is their frequency low? Do they feel emotionally stable during the conversation? Are they able to demonstrate their intelligence? All of this happens subconsciously. We can prep our appearance for a date or for an interview, but we can't change who we are without inner work.

How do you know when someone is an energy drain? Some of these points may seem obvious, but we'll go through them anyway.

You may have a longtime friend that you no longer resonate with. This can be difficult, especially if you've been friends since childhood. People grow and change over time, and sometimes people just grow apart. We make different choices. Our goals are different, and one day, you really don't resonate with that person anymore. Have a conversation with your friend. Ask them where they want to go in life. Sometimes this behavior can be detrimental such as addiction, or even apathy and not moving forward in their life. You may want to start your career, have a child, or travel. Your friend may be happy working at an entry-level position with no thoughts of moving up the ladder or changing anything about their life. If they are content, you have to let them be. It's their choice, and you move on.

Limit your time with draining people. We all go through seasons of life where one season we're on top of the world and the next we're wondering how we never noticed that our life was falling apart. We let too many weeds into our garden. Some seasons we feel as if we can do anything and become anything we want. We also have seasons of hardship, depression, loss. But these should be seasons and not permanent states of being. If you or someone you care about becomes stuck, reliving the past, you can be there for them; but you may need to create some space so you don't become suffocated.

Events in life affect us all differently. After illness or death or a child graduating and moving out, these emotions will process differently. What you believe plays a huge role in how you alchemize an event. If you believe there is a lesson to be learned or that this is not the only realm of existence for a soul to be in, you will alchemize positivity. This doesn't mean the situation is any less traumatic or the pain is lessened, it's that as you experience these feelings, you have a way to cope and process.

Those who do not have a positive believe system or a way to process traumatic experiences will think all is lost, that they have been a victim of someone else's actions, creating an emotional vacuum that is sucking them down. If you process this way, people may always seem to be trying to cheer you up. People invite you out, but you do not go. Some people start to avoid you, and your circle of friends becomes less and less. If this is happening, you need to find a way out of the dark. This can be done with eating from the light (quality fruits, vegetables and clean foods) and physically being out in the light outside in nature. Eat foods that are not processed and listen to uplifting/meditating music (music in the 400hz+), you can find this type of music for free on YouTube.

Your body stores emotions in your organs: anger in your liver, frustration in your head and neck, sadness in your heart, anxiety in your kidneys. Dis-ease is our body not being at peace and not being in the harmony with our environment. This is taking the intangible of the emotion and bringing it through the Ether and holding on to these emotions in our physical bodies. Since our energy body and physical body overlap, the energy becomes stuck causing blockages in the flow of energy. Shedding people who are dragging you down, even if its family or those you have known for a long time, will allow you the space you need to grow.

Intuitively feel your way through relationships and situations. If you embrace your intelligence, meaning you are learning from the world around you either by observation of others or through education and experience, your level of intellect on many different matters increases. Once you become aware of possible reactions of others, and can start to predict the most likely outcomes of certain situations, your intuition will also increase.

Concept 19
Evolution and Reinvention

Are you feeling in your gut that you're in the wrong place? I've explained previously how to rely on your gut instinct, pay attention to life cues, and identify what and who aligns with our goals. Now it's time to evolve and make choices we haven't before. Along the spectrum I laid out is skill building. Skills help you evolve and become someone new. Skill building could be traditional, but the traditional way of education is broken. So how do we learn? YouTube? Google? Okay, maybe. But what if we go back one hundred years, three hundred years ago. How did people learn? By doing, witnessing firsthand, training on-site, and apprenticeship. This is the way humans were designed to learn. Not being forced as a child to sit for hours on end, bound to a seat with threats of being in trouble if you didn't listen to a teacher who learned the curriculum to be regurgitated to the students.

What is your goal? It can be simply to move to a better part of town or move to the mountains, lake, beach, closer to family or move for love. Whatever the goal, what do you *want* it to be? In your ideal world, what would be the best outcome? We ask this question of our children: "What do you want to be?" "Who do you want to be?" "What do you want to be known for?"

Out of the answers comes, "I want to play outside," "I want to be with my friends," "I want to have fun." Sometimes you'll get, "I want to be a doctor or a veterinarian." Well, how do we accomplish what we want and still create enough energy to survive on? Any of these paths we choose has someone either already doing it, trying to do, it or trailblazing the path. Find your tribe. Collaborate. Work for someone who is doing what you want to do. Socialize with someone who has the lifestyle you want. Learn from others the way that worked for them. I can pretty much guarantee your exact path is not written in *one* book. It's written in *many* books, but your path is so unique you have to get the big picture before you can break it down for yourself.

You need to be in the right environment in order to evolve. Once you are, you will immediately start to see changes.

Reinvention—we have to let go of our old selves. Let go of old habits, old choices, thoughts, how we speak to ourselves and others. People wonder about manifestation. Fake it till you make it. Envision yourself having that thing, boat, car, house, or lover. Envision yourself skinnier. Um…No. The easiest way to explain it is, does a fit person eat chocolate cake? No. Don't eat it. *Make the choice of the person you want to be.* Make the choice of the fit person. Does a successful business owner sleep in? No. Wake up and get your day started. Does an attractive person speak harshly to others, complain about their food at a restaurant, and bitch about every little thing? No. If you recognize you're doing these things, change. That's probably why you're single. *If you want to manifest, you have to embody the habits of who you want to become.* Now how about manifesting a boat, car, or a house? Well, what type of person owns a boat, car, or house? What are the characteristics and habits of a person who owns those items? Do you have those habits, that mindset? Does a wealthy person say that "money is the root of all evil." No.

We'll stay on this for a moment because money is one of the three most requested manifestations (love and health are the others). Do you have a poverty mindset? As an example, say you run a business. You're making decent money, but if you paid someone $20 an hour for twenty hours a week to do the administrative work or tasks you don't want to you, it would free you up to generate another $1,000 in sales, possibly more. Do you know how many business owners do not take this first step to break out of the box to go to the next level? They have a fear of leveraging other people's skills to help propel them forward. Another example is a successful businessman wants to create a new way to purchase a product. He had most of the steps in place, a solid back office, people to sell, but no smooth mechanism to sell the product. Instead of spending the money to streamline sales, he continued the same process as it had been done for decades before. The refusal of reinvention shows a lack of connection with the Etheric Energy Exchange around him. When you don't take the steps to reinvent yourself, your capacity to take in more Etheric Energy remains limited.

Concept 20

Universal Laws vs. Programmed Truths, Half Truths, and Lies

Universal Laws are truths that withstand the test of time in our plane of existence and help explain how our reality functions. Truths have been hidden from us, distorted and programmed into us during school or inadvertently from our parents or other adults who were also told the same lies. Universal truths are there to be discovered by each individual person at the time in which they are meant to learn them. If you Google "Laws of Nature," there are seven:

1. Law of Attraction
2. Law of Action (polarity)
3. Law of Reciprocation (cause and effect)
4. Law of Rhythm (vibration)
5. Law of Reality (relativity)
6. Law of Masculine/Feminine (gender/gestation)
7. Law of Alchemy (perpetual transmutation of energy)

There are twelve Hermetic Universal Laws:

1. Law of Vibration
2. Law of Attraction
3. Law of Divine Oneness
4. Law of Compensation
5. Law of Polarity
6. Law of Correspondence
7. Law of Inspired Action
8. Law of Cause and Effect
9. Law of Relativity
10. Law of Gender

11. Law of Perpetual Transmutation of Energy

12. Law of Rhythm

We have discussed many of these without specifically describing the Laws. They are simple concepts, yet complex to understand and require open discussion. Funny how none of these Laws are taught in school or religions.

Through the bending and warping of these Laws, we are programmed to act counter to the natural process of these Laws. This is why abundance does not flow to us. This is why life seems hard or the roots of our creation seem to be counterintuitive, and we are required to go on faith. Just believe. You don't need facts and concrete reality, just faith. What truth would be required to lie and distort so much? For example, how is the Christian Trinity all masculine? Father, Son, and Holy Spirit? What happened to the Divine Feminine Mother? Law 6 of the Natural Law, Masculine/Feminine, and Hermetic Universal Law 3, 10, 11 and 12. If we stop and really ask ourselves, how could there be creation with *all* masculine energy? We know that the masculine alone cannot create life. Neither could the feminine alone create life. At least, not life as we know it.

We are trained to think that we need a good job where we work for a large company and are paid for our time. Trading time for money. This distorts Natural Law 1, 2, 3, 5, 7 and Hermetic Laws 2, 4, 5, 7, 8, 9, and 11. At said job, we are told how to work within the framework of the company. We are told what we can and cannot do. If a person is trained to be a doctor, they are trained in certain protocols that come down from the hierarchy of management of "best practices" instead of learning how the human body actually works from experience and learning how the body interacts as a whole, not just a foot, arm, nervous system, digestive system, etc. We are a whole person, yet medically we are dissected into parts and never receive whole, complete healing.

The distortion starts early by not discovering for ourselves our true skill sets. Our innate gifts are repressed or twisted, and we are programmed that trading time for money is the best use of our time and skills. This set up is not how human beings are supposed to live.

We can agree that those who win the wars write history from the winner's perspective. We never stop to question what the motives were behind those writings. We take proclaimed history books as truth, but is that really how the events took place?

The question becomes, Whose laws are you following? Hammurabi's, the Ten Commandments, social media's. We are being told how to feel about events by mainstream media, social media, our family, "respected people in our community" such as a pastor or local leader; but paying attention to external sources, instead of paying attention to our innate feelings, leads us to be out of alignment with our true self, our authentic self, and our beliefs become distorted. This creates an inability to use our intuition and decide what is best for us.

What is your personal assessment of what is going on in your world? You are not meant to take on the actions and feelings across the entire world. What is your world? What is going on around you that you can have impact on?

If you are putting out into the world something you know is a half-truth, a product that is faulty, a process with flaws, what does that do to the energy exchange? You receive what you put out. That is the Law of Attraction, the balance of the equation. If you are not true to yourself, if you are offering up a half-truth, a fake version of yourself, the Ether will act as accordingly and assumes that you agree with the lie, the deception, and your fake self. The Ether does not judge you in the way of right and wrong. The Ether only matches and reciprocates the energy you put forth. If your energy is a lie, it matches. If you are fake, it brings you what your fake energy is putting out. If you are selling bs, then you are bs. If the person on the other half of the equation believes you, yes, in the physical world they can be damaged, but in the Ether/spiritual realm, their karma will remain untouched. Yours would not. If you believe in yourself, your actions, your products and/or you believe you are doing the right thing, the Ether will reciprocate those beliefs. There is no hidden energy.

Spiritually, there is no intermediary between you and the Creator. No one is created in a higher status above you. There is no division among humans. If you really look at each spiritual dogma, the stories repeat. The flood story is in almost every religion, as is the concept of a savior and rules to govern the people. Horas (Osiris son), Attis, Mithra, Krishna, Dionysus, Jesus—all died, resurrected, performed miracles, and were born on December 25.

How is it that across continents, people seemingly worship the same God just known by a different name? No one can have power over you unless you give your power away. Your presence itself carries weight.

As much as we are in a rush to live our own lives, take time to notice others and care for other creations. If you saw a lost dog or cat on the side of the road, do you have time to stop and check on them? If you noticed a lost child in a mall or a park, would you stop and help? These are the Laws of Reciprocation, Vibration, and Alchemy.

If you feel like "I don't feel like I have a gift," you feel this way because your gift was taken from you. It was forced out of you when you doodled on your paper as a child and it was math time instead of art, when you were forced to stay quiet and listen instead of socializing with your friends, when you were forced to write a story instead of working with your hands to build your next greatest Lego creation. Rediscover your gift. Go back to the beginning of what you wanted to do as a child. What did you think was interesting?

Begin to explore what you like to do. If you like to paint, do it. If you want to go out and socialize with your friends, build those networks. Knowing people is valuable and being able to connect to people is a gift. If you want to travel, start by taking a different way home and explore the world around you. Ask to stay at a friend or family member's out of town if money is tight. Just by making those movements, the Universe will begin to rearrange the energy to provide more opportunities for new adventures.

These Laws are applied to each area of your life, yourself personally, your relationships, and how you receive energy by interacting with the world.

The Universal Laws affect all areas of our lives:

1) Personal

2) Family/relationships

3) Money and business

Begin to ask yourself how your feelings toward the world became warped, what those twisted truths are, and begin to untangle the Etheric Energy Exchange.

Concept 21
Poisoned Thoughts and Damaged Beliefs (Limiting Beliefs)

We live in a world full of social programming. It starts young with our parents and friends. Wanting to be accepted, approved of, liked, and loved. We learn subconsciously or through being told what we can and cannot do, what is or is not possible, what is socially acceptable and how to act to get the response we want. If we want love from our family, hopefully, it is a simple request: a hug or a kiss, a small child raising their arms to be picked up, crying because they're upset. For friends, it is an exchange of action, playtime, games, and, as we get older, excelling at academics, performance in sport, or physical appearance. The programming works by repetition and group acknowledgment of the standard. Societal pressure, peer pressure never ends.

Believing what has been represented as truth for so long is only natural. The belief that going to school and earning a degree from a prestigious institution will somehow equate to an enjoyable life. A job, doing what exactly? A medical degree taught by whom? Endorsed by whom? To what end?

We have also had our responses to stimuli preprogrammed. Once you notice the programming, you can't unsee it. The material the teachers are given to teach their students. All one way, all the same topics in history, all the same way to work a math equation. No new discoveries. News channels all running the same stories, using the exact same phrases. Your social media feed, all touting the same subject. You will notice the programming by the same use of words, in the same context from multiple people no matter their location. Certain words trigger a programmed response. Think it's a myth? Why then just before a storm does everyone go and buy toilet paper, bread, and milk? Are you going to pee more because it's colder out? The milk will go bad if the power goes out. At least, bread you can always find a use for, so that makes sense. Why do we not buy canned soup or vegetables?

When we are triggered emotionally, we either continue the outpouring of emotion because society is "for us" and supports that response, or we never share, because what if our response is beaten down in a public forum? What if we speak our truth and we are attacked? Do we keep speaking, or do we become silent? This is how the programming contorts your true self and controls your every move. No chains, no physical corralling you to do what you don't want. The program mold you in the way it sees fit to use your energy. Remember in a prior chapter? The energy exchange will balance. However, if you put protection around your energy, put parentheses ( ) around the equation if you will, you can protect yourself from the unseen negative drain. You can also protect yourself by calling out the negative that is bringing you down. Not by name-calling, cursing, being silent, or changing your stance but by acknowledging the truth in the situation. Hold fast to what it is you believe. Don't let people tell you the emperor has clothes when he is clearly naked. Don't let someone tell you something is good for you and good for your family when you know damn well it's not.

You should be able to share your truth in a peaceful and constructive way. A way in which your energy will be received by the other person and exchanged for energy from them. This also means that both parties are hearing and listening. Hearing is the vibration of their voice; listening is understanding the words. If one party is not ready to receive the other's message, a constructive conversation will not happen.

For example, my Husband tries to help if I am having difficulty painting the walls, using lawn equipment, or trying to open a can of salsa. He would just say, "Here let me help," and do what he thought needed to be done. But I didn't want his help. Could his actions make my life easier, yes, but I didn't want it. This goes for a child learning how to tie their shoes, completing their homework, navigating difficult conversations with a teacher at school. It may look like a good idea for the parent to step in and help, but ultimately, helping is hurting. No growth or independence happens when someone or something is constantly "helping" you.

Helping is also different from enabling. Enabling is providing your energy to someone else in a manner where the other person's container is leaking like a sieve. It's like giving money to a person who all they do is blow it on frivolous things or endeavors that do not bring value to their life for little to no return. They may hang out with people who drain their energy and money such as a bad relationship. They may go to the casino for excitement or drink or smoke it away. This enabling seems like helping, but it is hurting. Enabling is you standing in the way of someone else's life progress. It prevents them from learning the lessons they need to.

This can be incredibly harsh, painful, and gut-wrenching, especially when drugs and alcohol are involved. No one wants to watch their loved one be homeless or strung out. You can provide a safe space for them to come to. But giving money to help them out is never a good idea, because ultimately, money/energy alone is not going to solve their problem.

Paying for treatment is not a waste of money. Taking people in when they've tried leaving an abusive situation is not a waste of energy. You must fortify your own energy by putting a bubble around your emotions, a parenthesis around the energy equation, and not letting them take more than you are willing to give. Yes, be there for your child, your best friend, your sibling but do so with the understanding of their position. Are they willing to receive the help *and* alchemize their change? What are you willing to lose to help them?

How do we receive information that is being presented to us as truth? Ask question and research as much as you can on the topic. How is this possible that a positive message becomes inverted? When your intuition becomes activated, you will start to spot when you're being manipulated.

How do we prevent the leaking of energy? By changing our damaged belief system and understanding how our thoughts became poisoned in the first place.

Michael Bernoff, a neurolinguistic programming expert, says, "Asking why is insanity. Asking how we got here is progress."

We were taught what to think, not how to think. Plato must be rolling over in his grave watching how low humans' mental capacity has all but dissolved. That is, if Plato was an actual person and not a concept of a teacher in the first place or a made-up pen name. How do we shake loose the old programming? We begin by sitting with ourselves. Meditation is not clearing the mind. It is questioning the mind/Ether and waiting for a response. How is it that I chose this crappy job? Wait, listen, and process. Let your mind wander. Eventually, your mind will focus. It will forget about what you have to do later, the dirty dishes or if you forget to brush your teeth. Ask the same question again or ask another in a succession of what is bothering you.

How is it that I chose to live here? *I've always lived here or it's what I can afford.*

How is it that I keep living paycheck to paycheck? Possible response. *My crappy job.*

How is it that I chose this crappy job? *It was the only one available, or it was the only one I was qualified for.*

How can I change? *Bingo.*

You will eventually get a response from a question that will require action. Listening and taking action on this question is how you begin to change your life.

Wait and listen. Your mind will begin to interact with the Ether and respond. This is the true beginning of meditation, the waiting for the response from your higher self. The response is not clouds opening and a single ray of sunshine beaming down on your cross-legged body, which by now, be honest, your legs have fallen asleep and you have to pee.

This is the start of your ideas/thoughts/inspiration of what you can choose to do differently and *how* you can accomplish these tasks.

Getting to the point of true meditation and question-answering will get faster. You'll also be able to do it while not in a cramped, I-don't-practice-yoga position. You'll get it while doing mundane tasks such as the dishes, vacuuming, mowing the lawn, driving the car (be careful with this one or you'll end up an hour in the opposite direction of where you need to be), or while in the bathroom. Guys, this one is probably for you since you spend so much time on the porcelain throne anyway (legs falling asleep is still a common occurrence).

Also be wary of the limiting beliefs. "It was the only one I was qualified for" and "I've always lived here." Words like "only," "always," "never," "because," "I can't," "I don't," carry limiting beliefs. You can start by asking "How did I come to think this _____." Not why, but *how* did I come to think that one book written by man contains all the beliefs necessary for me to process the complexity of the entire Universe. How did I come to think that not eating carbs is best to lose weight? How did I come to the conclusion that I think that my spouse/significant other thinks I'm wasting my time on my dream? How is the other person acting to make me feel this way? How am I responding to my loved one putting me down? How do I know my path is the right one? Etcetera, etcetera.

Start at five minutes, either first thing in the morning or right before you fall asleep. If you do this before bed and you are a naturally intuitive person, you will start to get answers in your dream state (brain waves: beta, alpha, delta, theta, gamma). Make sure to keep a dream journal. Write down everything you can remember as soon as you can, to not lose any details.

Interpreting dreams is a whole 'nother thing, and that is not something I'm good at. Also, scenes in your dreams will have a particular meaning to you rather than having a third party interpret them, so try and decipher by researching dreams yourself. Research in this case means gathering many opinions and taking what resonates with you.

As a note to the naturally intuitive (and if you're working on it), there is something called sidereal time. You can find an app for it on your phone. You put in your location, and it will give you your sidereal time (https://earthsky.org/6/10/2012). Christopher Crockett says, "A sidereal day measures the rotation of the Earth relative to the stars rather than the sun…A solar day is how long it takes the earth to rotate once—and then some. A sidereal day is 23hrs 56 min 1/4 sec. is the actual amount of time needed to complete one rotation. In this system the stars always appear at the same place in the sky at the same time each sidereal day. Sidereal noon is when the vernal equinox—where the sun sits in the sky at the first moment of the northern hemisphere, Spring— passes directly overhead."

David Wilcock mentions the optimal time for ESP is 13:30 LST. Fifteen minutes before (13:15 LST) through fifteen minutes after (13:45) is optimal heightened time to mediate.

Lost knowledge of how our bodies work and interact with the Ether/natural world and Universe will hopefully begin to permeate the psyche of humanity. The more we learn from ourselves by asking ourselves questions, we will begin to match the natural vibrations around our bodies, pulling in good things and naturally repelling what does not resonate. So much so that evil cannot even come into our field of thought. Negative thoughts will stop. Toxic people will be turned away upon feeling they cannot penetrate our "bubble." Our "aura," our personality, demeanor, our energy will repel them. These are all words meaning the electromagnetism of our vibratory field generated by and conducted through our physical hearts will sustain our positive being and increase our frequency (ascension).

Physically, how do we help our bodies release the negative and prepare as best we can for meditation. Taking salt baths with Epsom salt or Himalayan salts. Using crystals, conductive stones or metals such as silver or gold for physical and psychological properties. If you don't believe in crystals, stop using your phones and computers now. Both contain precious metals and quartz crystals. If you don't believe in energy manifestation through metals, go live with the Amish because it's copper wire running electricity though your walls. Crystals are not magic; they're science. True, real science, not what your programming has taught you to think about them, with your knee-jerk reaction to the Etheric.

An example that is not mainstream of how energy is contained in substances is limestone. Limestone is thought to contain emotional energy. Energy that can be imprinted on an old house for example. As your electromagnetic field moves through limestone, energy can be released in the form of a residual haunting, repetition of words, and even scents can be played back with this interaction. Much like playing back a cassette on a tape player (ah, eighties reference). This playback has also been found in Egyptian pottery. Voices of the Egyptians were recorded as they were carving grooves into their pottery. The clay became conductive, a medium of recording energy and entrapping it in the clay, like a vinyl record.

The use of silver in your left hand and gold in your right also helps to conduct energy through your body when meditating. This energy pushes and pulls the chi through your body in a safe and clearing way.

Scents are very powerful. Scents are the number one memory trigger. Scents of coffee, bacon, or cookies can be comforting and remind us of home. Pine, frankincense, and apples remind us of the holidays. Pipes remind me of my grandfather. Honeysuckle, morning glories, and pink roses remind me of the Ether/spirit realm. As you explore scents, you will find one or a few that help you get into a relaxed state. By creating a sacred space in our home, where we clear the energy in the room and create a space to just be, allows for our frequency to become pure (not agitated by outside stimulus).

Physical health is also important. There is a reason the adult human body is up to 70 percent water. It is how we flush toxins from our bodies, conduct electricity, and interact with the world and Ether around us. When our bodies become polluted with chemicals, the interaction between our physical being and Etherical body becomes disrupted. It's like trying to send electricity through bread. Not very conductive.

Another way to think of cleansing out bodies, and I have to be honest, this visual always gets me to drink more water. I can't remember where I heard this from, but its good. Your body is like a toilet. The waste flows out through flushing with clean water. If you put "crap" in the top of the tank (sodas, energy drinks, processed foods), you will not have a clean body. There will be perpetual "crap" in your system. These toxins store in your organs. When these toxins store in your organs, they become agitated, and when your feelings go to store in these organs, it creates an even more toxic reaction.

When our bodies are hanging on to too many toxins, bacteria, or emotions we begin to feel sick and serious diseases and parasites such as cancer invades the body. What do we go to? Natural or synthetic. Pharmaceutical companies are not able to "patent nature." Meaning, no profit. Interesting how the main ingredient for Tamiflu for instance is star anise. Pine needles also carry a similar effect. But if you mention these natural remedies to people who have never studied plants, suddenly, it's taboo or you're a witch doctor. You can look it up yourself. The information is buried, but not hidden completely. Star anise contains shikimic acid, a plant passed compound that is the precursor to oseltamivir. (And yes, I had to look that up. I'm not a doctor and don't pretend to be, but I do know bullshit when I see it and avoid stepping in it the best I can.) The difference in the synthetic product and the natural is exactly that. Synthetic chemical additives that change the natural product into something your body doesn't even recognize, causing more stress on your liver and kidneys to filter out the poison and keep what was good, natural, and recognizable by your body.

How is it that weeds are weeds and bad for us and we shouldn't have them in our yards? How is it that no one knows how to grow food anymore, let alone medicine? I'm not going to answer that here, but please ask yourself and see what you get. How is it that I think my lawn looks good with no life in it except Bermuda grass, which isn't edible and serves no purpose?

We're provided a choice of foods, and we go by what tastes good. Are we toddlers? We do no research on our own. There are no nutrition classes taught in any school in the US for a reason, a programmed reason. How is it we treat our bodies the way we do? Little sleep, crappy food, bad medicine. We're programmed for self-destruction. Work harder, not smarter. If you work hard, you'll get a good job and get paid by someone else who determines your worth. Do you choose to stay up late and ignore your body when its tired? Here, we need to separate the mind and body. Our body is our avatar. It is not who we are but what our soul is choosing to use to interact with our environment.

The disconnection of the mind and body causes us to be unaware that we are causing harm to one or the other. We stay up late. We go ahead and eat the meal that causes indigestion instead of making a different, healthier choice. Why? Insanity. Is the convenience and habit of certain foods that great? Women wear uncomfortable shoes to look cute/sexy/stylish. Pants that are too small because we refuse to acknowledge the size we are. We fill our heads with shows, videos, and apps that steal our time in order for us to "relax" and *disconnect or unplug* from the world. If we don't start paying attention to what our minds and bodies are exposed to, your body will be screaming with disease, inflammation, irritation, and imbalance, which causes you to reach again for something external in the form of a pill, addiction, or something else that's bad. It's spiraling out of control. An environment that has been perfectly programmed since your birth to destroy your soul; to separate you from your Creator; to deafen your senses and turn you into a zombie, ripe for programming and control. All because you gave your power over to something else. You reached outside of yourself, to religion, to school, to the government, to the acceptance of social-media friends. Stop. Just stop. Breath and go within. Ask yourself questions, one at a time, and use this to disconnect from fakery pretending to be reality.

This is where the Metaverse comes in. Is the Metaverse an alternate reality? Is it harmful to the psyche of the physical, incarnate human being? Does it create distortions, disassociation disorders, and confusion as to what is real and what is fake? Wait, are we already in a Metaverse?

Online gaming with real physical friends *is* an alternate reality. We are mentally/Etherically real, physically real, simultaneously existing in two realms. The invention of video games makes it much easier to convey timelines, worlds, realms, and the existence of the incarnate/body/physical world simultaneously existing and interacting with a metaphysical realm. It is the separation of mind and body, astral projection at its finest with help from technology. You never thought about it that way did you? TV was an easy first step; but interacting with others without your body happens all the time, over the phone, or through a video game. Pretty trippy and you didn't even realize what was going on. Phones, the Internet, online gaming, communicating over long distances, being in a different world—these are all things our bodies can accomplish without the use of electronics or AI technology. Technology is all based on what our DNA is actually capable of; however, if we are not controlled, we become too powerful, too independent, and no creation can restrain us except our Source. This is why Lucifer was afraid of humans. Lucifer envied us and detested us. I use the past tense, because an event has occurred that makes the control past tense.

So how do we stop the control? The transition is slow. Deprogramming and reprogramming takes time. Here are a few ideas:

- Instead of watching TV, go outside.

- Instead of video games, go out to dinner with friends.

- Instead of making purchases online, go into a local store. (This is part of why malls were disassembled. Too many people were meeting their friends there to socialize and mingle with each other in the physical world creating too much physical attachment.)

- Instead of online dating, meet people face-to-face.

- Stop buying food online or even in a big-box grocery store. Find local merchants and farmers markets whenever you can.

- Go outside and meet your neighbors.

All of our world aspects have moved into the Ether, where the thought of a shoe can be purchased and manifested into reality. Is this the same Ether that our souls, ideas, and dreams exist in? Yes and no. Yes, because ideas, goals, creation of sales, exchange of energy of money occurs in the Ether before the physical. But no, because AI has simulated the Ether and is incompassed in a closed environment. AI must create a replica of life and reality in order to create a new realm or construct. AI creating the internet and the metaverse is technology trying to play Creator. A copy of the truth.

Before the year 2000, people purchased what they needed 100 percent in the physical realm.

After 2000, people buy in the Ether to produce it on their front porch in the physical.

The transaction is the same, you are the same, the energy is the same. *Where* the energy exchange happens is different.

Money is slowly becoming Etheric, or not physical. Other than the dollar amount you see on your computer screen, you never physically touched your money. You didn't physically see the product. You believed it was real, and then a couple of days later, your belief turned into a physical product.

Is the Metaverse going to lead us to a world where AI tries to take over our minds and manifest itself into reality? I hope not. There is duality in all things—good, evil. It depends on how we use our energy. That will be the difference. Manifest abundance, joy, and peace; and it will be. What your mind thinks and what your heart/soul believes becomes physical reality. Universal law, the energy exchange, has to be completed. The equation has to be equal—that is the science of spirituality.

Sharing your thoughts and beliefs with the world can be a scary endeavor. You may think you have the most important message ever that will save everyone! But if people are not willing to receive it, it doesn't matter. Why do you believe what you believe? Who told you those facts? Who told you "the science"? Did you experience any of it for yourself? Remember is it the winners of war that get to determine what is fact and what is erased.

This is why it is so important to go within. The distortions of truth to get you to use your free will to buy into a preprogramed, corrupt system. Please know that Universal Law and karma know that deception is not free will. This is why changes are happening. No matter how warped DNA becomes, after a few cycles of letting nature replenish itself, creation will go back to how Source intended.

There are studies of insects that were genetically modified: two heads, extra sets of wings, extra or no antena. The changes were not due to adaptations to nature, but through man's genetic manipulation. After six generations of being left to breed on their own, they returned to normal form. To their original template, or their original natural program. The human body fully regenerates every seven to eight years. Every muscle, organ, and tissue regenerates. We have to cleanse ourselves of the poison and chemicals from our environment. Start discerning right from wrong. Learn the truth of existence through actual science, organic science, not mainstream, made-up, fact-checked-by- independent-paid-for-fact-checkers science. For example do you know an entire dinosaur skeleton has never been found? So what we see are total speculations as to how dinosaurs looked. Do you know how many bodies of giants have been thrown into the ocean by the Smithsonian to destroy evidence contrary to Darwinism, which is only a theory anyway?

What does it mean to be triggered? There are a series of phrases, words, constructs, social taboos, spells if you will, that when a phrase is spoken or pictures are shown of a person or scene, it immediately triggers a response. My goal here is not to push those triggers but to help you get to the root of why certain phrases trigger you and how to move past those initial gut reactions to a place of understanding yourself, others, and practicing loving detachment. Obviously, this will not miraculously happen in the duration of one chapter of a book, but you will at least get an opportunity to sit with yourself and ask how is it that you came to your stance on certain topics.

I'm going to give you a series of words. We're not going to discuss them. I want you to *sit* with the emotion and ask yourself *how* is it that I came to react this way? Not *why*, but *how* did I come to this emotion, conclusion, this belief? That is all. It sounds simple, but this is your shadow work. When a word triggers you, stop on that word, and work through the emotion. This paragraph of words may take the rest of your life to untangle.

God. Satan. Abortion. Mass shooting. The Church. The Vatican. Religion. Priests. Sex. Government. Republican. Democrat. Black. White. Mexican. Asian. Special needs. Learning disabilities. Gay. Straight. Women's rights. Black Lives Matter. All Lives Matter. Guns. Entitlement. Spirituality. Herbalism. Mysticism. Tarot. Flat Earth. Round Earth, the Bible, the Koran, Jesus, Muhamad, Lucifer.

Sit with the emotion. How did you come to hold the beliefs you have? Were you told what to think? Were you programmed? What is your own, personal experience? Not what you saw on TV, not in a book you read, or what heard from others. What is your own experience? What is *your* truth? Are you living someone else's beliefs? How do you know what is true? Is what you believe in the highest and best good? Is your belief in service to others?

If you breezed through this exercise or if you don't want to take on something that heavy right now, here's a trigger I always found funny. The weather app has indicated its going to snow tomorrow. Your work will be closed, and schools will be closed. What was your programmed response? Was it to go immediately to the grocery store for bread, milk, eggs, and toilet paper? Why? Why is toilet paper so important? Does snow make you have to go to the bathroom more? Why not water or canned soup? Why is the program not to head to the gas station? At least add some maple syrup to the list so you can make some French toast on your snow day.

You know you are having a programmed response when your reaction is automatic but in no way helps your situation. For example, someone says they're a Democrat and you're a Republican. Instinctually, you think that person is a dumbass. Or you're a Democrat and they're a Republican and you automatically think that person is a dumbass. You know you're programmed when you cannot hold a conversation with a differing viewpoint and not be angry. You use words like "lazy," "chauvinist," "fascist" and go straight to labels. Labels are a hardwire program that are command words to get the program to run. I'll say that again. *There are trigger words being used to get a program to run.*

Practice bringing up topics and remaining neutral. Try not to have a feeling one way or the other. If your emotion rises, acknowledge how you feel and still listen to the other person. Sit with the opposite point of view without judgment. Reevaluate how you came to your truth. Did anything change? Did you learn a new fact? Having the ability to explore ideas from a different perspective either generates new knowledge or further solidifies how you feel. Have these interactions with others without absorbing their negativity, without judgement, without anger. This will allow you to let in all aspects and come to your own conclusion. Knowing full well that with additional information or experiences, it is okay for your stance to change or stay the same on any and all topics. It is okay to change your mind. It is okay to be angry when you found out you've been lied to or programmed.

Many times we don't even want to entertain "the other side" or "their beliefs." Which side are you on? The truth is, sides change, opinions change, as they should when we grow and evolve. As you acquire more insight, use your own intuition and firsthand experience as much as possible. Don't be programmed.

Concept 22
Spiritual Entanglement

What lessons did you come here to learn? This is a difficult one. I am not saying anyone ever deserves a situation they are placed in that they have no control over, especially children. My friend was at a point in her life where she felt stagnant (actually drifting more into the negative), totally unhappy, unfulfilled; but somehow, she still had a spark of life inside of her. Going back to synchronicity and repetition, the Universe will keep placing the same lesson before you until you pass your test. In my friend's life, it was narcissistic and abusive men. From the time she was a child, she was sexually and physically assaulted by a man who was supposed to love, nurture, and protect her. And no one saved her. Not her mother, not her sisters. They did not even acknowledge what was going on. So my friend endured the best she could. She created walls, a separation from that abuse from the rest of her life, and the first chance she got, she was out of that living hell.

She became a responsible young woman (I still am surprised how she isn't strung out on drugs or committed suicide). She married and had children, but little did her conscious mind know, that her subconscious was stabbing her in the back. Her husband began to verbally abuse her, at first little things. "Why can't you do this right? "Why isn't the house clean? Don't you love me?" "Why don't you make enough money? Why don't you try harder?" This time she was an adult. She had to strategize because she had two small children, no job, and no safe place to run. She had to rely on herself. She got a job, saved some money, and left. (As a brief fact, statistically, it takes at least eight times for someone, usually a woman, to leave an abusive relationship. Also 80 percent of the time after leaving, the abused is murdered by their ex. It is dangerous to stay, and it dangerous to leave.)

Her hell continued as her ex told her outside the courtroom that she would never have her children. The judge gave favor to the ex (I hope they rot in hell); and the two children, a boy and a girl, were given to someone who did not love them, couldn't care less about them, and physically abused them daily. (The boy was abused with objects from the home, from spatulas to hammers, to cables. He was less than six years old.)

My friend was finally able, after years of fighting to get her children back, but the damage was done. Two more human beings exposed to hell. Two more people becoming fragmented in their mind and soul. My friend went on to have another marriage and one more child. Again, choosing a selfish person, but at least not abusive and only slightly narcissistic.

So why the repetition? Why keep facing one tragic situation after another? How would someone not give up or go insane or end up in jail for taking justice into their own hands?

The more evolved your soul becomes, the more difficult your trials become. A weak soul will never choose to learn the lesson to learn how to recognize manipulative, dangerous people; to learn the lessons of understanding what it feels like to be helpless, sad, and angry and then figure a way out.

Why would any soul choose to have this experience? Maybe because you wanted to give the other soul a chance at redemption? A chance to work off karma or to make the right choice this time around? Well, three men failed their tests. Eventually, their souls will reincarnate and have a chance again to learn, understand, and grow, or potentially being placed in a state where they need to reflect and learn before any more incarnations can be made. People who are evil may also be devolved on the soul evolutionary track.

Making it to the next level of enlightenment is what life is about. Life is a physical classroom for metaphysical souls to experience and learn from each other; to learn more about themselves and how energy acts, interacts, responds, repels, positive, negative—the push and pull of the Universe.

I will take a moment to briefly describe hell. Her life sounded like hell, did it not? Do the men who inflicted violence and damaging experiences deserve hell? When we leave out bodies, hell is a state of being. It is said that hell is a prison with a door that is locked from the inside. My friend could have chosen to stay in hell, but she got out. The men are locked inside. They do not have enough perception of themselves to stop damaging others and therefore their own souls. It can take many lifetimes to learn to defend yourself from abuse and set healthy boundaries. It also takes many lifetimes to learn not to be the abuser.

Let's learn about manifesting a good experience. Another friend, after shedding a partnership that no longer served him, chose a career path using his intuition through the meditation methods I described. He took himself from a life of mediocracy, accepting what was placed before him, to attaining love and business connections through his skill and passion, and created money (energy). By following his intuition, he made choices in his highest and best good, but also the choices he felt led to make that lined up with our Creator and Universal Laws. Abundance spilled over into every aspect and doors opened, but he carefully and intuitively chose which ones to walk through. People were attracted to him. You will notice, once your vibration changes, people will want to be near you. It's like a plant toward the sun. This time you are the sun sharing your energy with others. Some energy is given freely by the overflowing of abundance, kindness, generosity, and being a good human being. Other energy is more focused. He chose to focus his energy on people and relationships that would bear fruit. Through discernment and experience from the school of hard knocks and learning from poor choices, he is now able to not waste time by quickly identifying what is a good fit from the beginning.

By doing your own inner work, when you are clear in who you are and what you want, and send a clear message to the Ether to attract those specific criteria, the Ether can clearly fulfill the transaction. That criterion has a specific vibration. If you are looking for a relationship, that vibration doesn't come with a specific body type, hair color, or accent. It does come with how you want to be treated, what you like to do; the vibrations of what you find interesting; and the resonance of the same likes and dislikes. How you present yourself to the world and knowing your identity will clarify for all others how they should interact with you. When you walk into a room of peers, friends, or family, what are you know for? Who are you to them? If you are not able to clearly understand how others perceive you, you will not be able to move forward in life on a clear path or attract what you want.

Spiritual entanglement is the similar if not the same as quantum entanglement, meaning that particles that have the same frequency can resonate together no matter the time, distance, or space (realm); meaning that communication travels instantaneously and any stimulus can have an effect on both particles.

Concept 23
Types of Energy Bodies and Brainwaves

Your energy body is a combination of your place of birth, date of birth, the energy that surrounded your birth at that specific time space and space time, as well as the astrology and your chakra system. This is all encompassing of a human body. This type of spiritual analysis of the human body was developed in 1987 and is definitely not mainstream. (I should say the rediscovery of the esoteric knowledge happened in 1987.) If it was, everyone would have a chart done from birth to help guide them in the best possible way for success. This chart guides you to what your purpose in life is and how best to accomplish your mission.

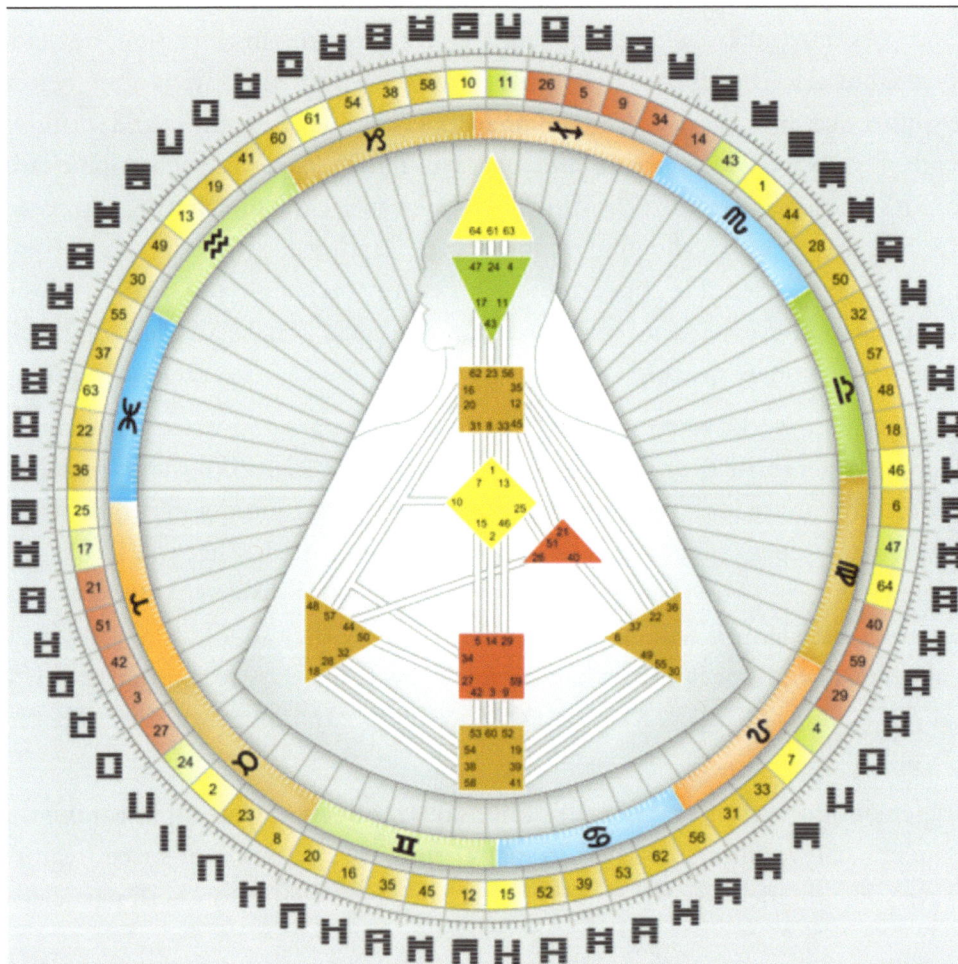

There are five energy body types: Manifestor, Manifesting Generator, Generator, Projector and Reflector.

The first three are energy beings. They can create their own energy to manifest their dreams.

Manifestor: 8–10 percent of the population. They are the initiative takers, the leaders, innovators, and creators of change. Examples are managers, inventors, and business owners.

Manifesting Generator: 32–35 percent of the population. They put their creation out into the world. They channel their ideas from the Ether and bring them into reality. Examples, entrepreneurs, producers, manufacturing, farming, providing energy for other people to use.

Generator: 35–38 percent of the population. They provide energy and guidance for supporting others on their journey. They succeed by doing what they love to do. Examples, teachers, trainers, in the area of study or profession they love.

The fourth and fifth type are non-energy beings, meaning they draw on the energy of others to help complete their life's mission.

Projector: 20–24 percent of the population. They direct others and provide guidance. They take a larger outlook on the collective goal. Examples are project managers, CEOs, community leaders.

Reflector: 1–2 percent of the population. They help to stabilize the emotions and energy of the collective. They observe the world around them. They astrologically are most affected by the cycles of the moon. Examples are professional musicians and writers.

Your energy type helps you to understand how to exchange energy with the world around you.

Your authority is how to know if a decision is right for you, it is your inner guidance, your discernment.

Your energy body, in conjunction with your physical state of consciousness, creates the ultimate spiritual embodiment. This is your spirit and body firing on all circuits. Your soul being fully connected to the physical vessel.

The different type of brainwaves in increasing vibration are delta, theta, alpha, beta, and gamma. Each brainwave phase has its function on how your energy body is interacting with the physical world.

Delta: .5–4Hz, deep sleep

Theta: 4–8Hz, sleep or daydreaming, meditative state

Alpha: 8–12 Hz, conscious brainwave, relaxed

Beta: 12–38Hz, high awake, alert and thinking, working in the physical world

Gamma: 35hz and above is a high state of concentration

It is thought that the theta brainwave state is the most connected to the Ether. You are in a relaxed state of being, but not fully conscious. It's the head space you're in when you first wake up; still feeling a bit groggy, not fully awake or aware of your physical surroundings. Maybe you're waking up from a dream. This is the state that hypnotherapy takes you to, to work on your inner knowing or correct behaviors/program at a subconscious level. Going from an active meditation (alpha) to relaxing meditation (theta) is the state of being we are ultimately trying to get to pull our ideas from the Ether. This can be done through prayer, listening to soft music, being in a relaxed position, with no disturbances.

In order for all of this energy to pass through the circuits of your body in the most efficient way possible, we must not eat sugars or chemicals. No processed foods. We need to drink clean water and eat from the light: veggies, fruits, and anything with one ingredient. We need to eat food that is as unaltered from its natural state as possible.

Concept 24
Protecting Your Energy

There are many factors that go in to protecting our energy: what we eat, how much we sleep, and removing stressors from our lives. In this section, we will discuss a bit about how we can protect ourselves from unnecessary draining of energy and some concepts to think about as we go about our lives.

Setting boundaries: It's okay to say no. Saying no when you are tired and do not have the energy to freely share with others is one of the best ways to set our boundaries. Protecting our energy from energy suckers, such as a draining family member, friend, or work situation, takes practice. From childhood, we are told to do things we do not want to do: brush our hair, take a bath, eat our vegetables. As we get older, we may be too tired to work out or share time with our children or spouse, but this is important. Even when we're tired. But this is not what I'm talking about. I'm talking about the abuse of time. For example, you are a salaried employee and you do not get paid for working overtime, yet your boss continues to request tasks of you because he/she will know you can't say no. You became the "yes" man. No one wants or should be the go-to person that takes on the brunt of everyone else's lack of work. When we're in our young adulthood, our friends want us to hang out all the time, party on the weekends and perhaps get involved in some extracurricular activities that are not so healthy for our bodies. It's okay to take a break and rest. It's okay to be a hermit for the weekend and sit on the couch (after your workout).

Taking care of our physical body: Eat as cleanly as possible. Eat as close to one ingredient and the whole food, fruit, vegetable as possible. We can stress eat and use food as a calming mechanism. This is my biggest issue. Changing the action of eating to exercise for calming is one of the best ways to change our habits and use our energy to help our body. Exercising in nature or having a workout partner are also good ways to stay motivated and accountable for our health. Take salt baths and use products on your body that are not toxic. If you bought your cleaning or body products from a chain store, they're most likely contain energy-disrupting chemicals that are detrimental to your health.

Take care of you mental body: Rest, take time to relax, and meditate. Do what makes you feel happy, listen to music, listen to positive meditative mantras.

There are no secrets in the Universe. You just haven't found the answers yet. To change the world, you have to go out into the world. Do not think that you are so significant that God couldn't do it without you. Do not think you are so insignificant that God cannot use you to change the world.

David Wilcock says, "Hope is an acknowledging of an underlying fear. In order to ascend, you must eliminate fear and the necessity for hope." (This is the power of belief. Believe and know that you are on the right path.)

As a someone who holds the Christ light, it is not your job to convince people about who the Creator is. Your job is to show people love, compassion, understanding, tolerance, grace, and giving.

When obedience becomes your first thought for survival, you have lost your humanity. It doesn't matter how much you expose the evil of this world if you do not expose the darkness within yourself.

Most people will go along with anything if there's enough fear.

Ascension doesn't happen to you; it comes from within you.

You cannot sacrifice your own personal truth for someone else's.

Ladder of Belief and Prosperity

Fear is an inhibitor. If I'm being inhibited, what am I afraid of? What you are afraid of is what you are not doing that you should be doing. This is the *hidden* thing you're afraid of. Are you afraid of being seen? Making sales calls? Being seen as stupid? Are you afraid of rejection? Uncovering your hidden fears is the greatest awareness to help you gain your freedom.

Belief	Belief in a Higher Source; belief in yourself; you have the ability to repel negative thoughts and beliefs before they enter your field; confidence in yourself; confidence in your knowing; you cannot be swayed from your truth, you can entertain an idea even if you do not believe in the idea presented.
Faith	Knowing what you have to do to "be good" or have a happier life, able to identify negative thoughts and push them away, follow the rules to get to a certain outcome.
Hope	See what you want and have a desire to get it, but you still entertain negative thoughts; hoping for the best; lacking confidence in yourself, others, or a Creator's power.
Fear	No thoughts of your own, you are controlled by your emotions, easily swayed in what you think and how you feel. Fear keeps people in a place they don't want to be because they're more afraid of having less than experiencing the happiness of having more

I can't remember if I channeled this or heard it from someone else's training. If you know, let me know, and I can give them credit.

- Premature transparency—you want people to feel bad for you, you share your issues, not getting better, just complaining about your situation.

- Immature transparency—you are aware of what you lack and hide your insecurity before someone else finds out, you have no intention of fixing your identified fears. Share of yourself and skill only to compare yourself to others. Am I better than they are?

- Mature transparency—this is where the acknowledgment of being broken is communicated and invite accountability to become better.

Trust a teacher who's been through some shit. Push through the depth of the pain. Testing through perseverance. God/Creator/Source loves you and knows what you're going to face. "You won't understand, I have to prepare you with strength and perseverance, consider it a joy to face trials, you know that it will develop in you the perseverance you're going to need and not lack anything" (Book of James).

God loves you too much to answer your prayer at any other time than the right time (Hebrews 1:14–18).

People see what they're conditioned to see.

Leave people in the chains they're content to wear.

Creator gives you what you need, not what you want. Don't confuse your wants with your needs. Be open and accept gifts in whatever form they come in.

I don't love you for what you do; I love you for who you are.

Sometimes it's difficult to tell if you're watering a seed or a dead plant.

While in pursuit of the life you want, you still have to live the life you currently have.

As long as a slave has the perception of freedom and independence you can bend them to your will with ease (*Meredith Peasey*).

Protect your energy. No one else can do that for you. Do the shadow work, explore your triggers and work through them. Begin to understand who you are, what you are, and why you decided to be on Earth, at this time. Use the Etheric Energy around you to lead the life you want.

Credits/Sources

Start the Day with Chakra-Balancing Shower Meditation. (Elephant Journal)

Moving heavy objects through vibration: https://youtu.be/-2YXX7WXNKY?si=Rb3XECVL7qipeFG1

Experiments with sound: https://youtu.be/rYrdiQckGhw?si=tqiOrMdgmitADJfj

Water and sound: https://youtu.be/uENITui5_jU?si=ysByGXQoYlGsECLx

Ancient books: https://spiritual-minds.com/religion/Gnosticts/Essene%20Gospel%20of%20Peace1.pdf

Energy Body: www.myhumandesign.com

Emotional Body: www.humansystems.co.

Suggestions to open your mind to other possibilities:

Rumble

Amazing Polly

Beyond Mystic

ECETI Talk Radio

Suspectsky

Catherine Edwards

Esoteric Atlanta

YouTube

David Wilcock books and early presentations

Suspicious Observers

Dutchsinse

Breathwork Beats

Dr. Sharnel True TV

Earth Star Academy

Eckhart Tolle

Jarid Boosters

Mind Unveiled

The Human Diet (https://youtu.be/p2YOo1kJV74?si=gv-b_lNlqy11ONyb)

MrMBB333

Peter Maxwell Slattery

Sadhguru

Stefan Burns

Taylor Moon

Praveen Mohan

The Why Files

Beer Biceps

DNA Awakening

Herbal Jedi

Maison Jupiter

Music for Body and Spirit—Meditation Music

PK Costello—History of A Lost Earth, What on Earth Happened

Spooky2 Rife

Realm of Astrology

Rise.tv

TarotbyJanine.com

Beyondmystic.net

The 5th Kind

Dr. Masaru Emoto's Study of Water

www.ingramcontent.com/pod-product-compliance
Lightning Source LLC
Chambersburg PA
CBHW041448210326
41599CB00004B/175